International Halley Watch Amateur Observers' Manual for Scientific Comet Studies

Part I. Methods

Stephen J. Edberg

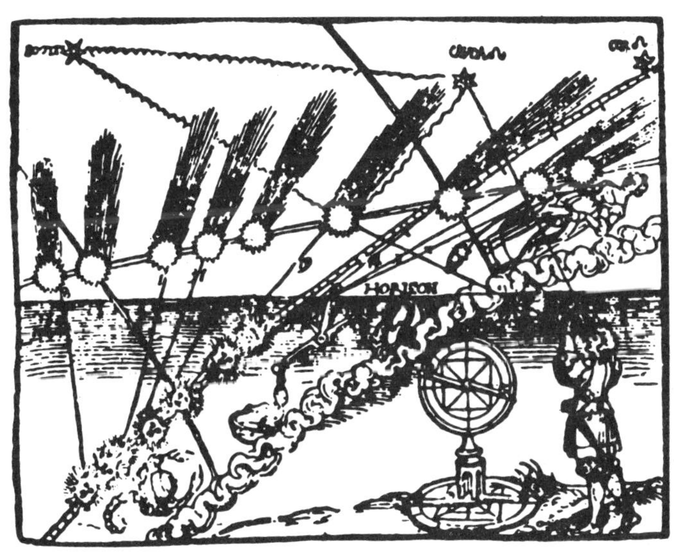

The research described in this manual was performed at the Jet Propulsion Laboratory, California Institute of Technology, under contract with the National Aeronautics and Space Administration.

Originally published by the National Aeronautics and Space Administration
and by the Jet Propulsion Laboratory of the California Institute of Technology

March 1, 1983.

This edition published by

Enslow Publishers
Bloy Street and Ramsey Avenue
Box 777
Hillside, NJ 07205

and by

Sky Publishing Corporation
49 Bay State Road
Cambridge, Mass. 02238-1290

November 1, 1983.

Cut Here

OBSERVER INDEX

Please tear out this form and fill it in as completely as possible if you plan to submit observations to the IHW. Also fill in the duplicate in Part II for your own records. It is important to read Sections 2 and 4 and the section describing your area of participation in Part I of this manual before submitting this index form. Return this form to Stephen Edberg (Jet Propulsion Laboratory, 4800 Oak Grove Dr., T-1166, Pasadena, California 91109, USA).

Name ______________________________________   Telephone:
     (Last, First)

Mailing Address ____________________________   Day ________________________
                                                     area code + number

____

____   Night ______________________
                                                     area code + number

____

Areas of Participation: (check all that apply)

____ Visual Observations   ____ Spectroscopic Observations
____ Photography   ____ Photoelectric Photometry
____ Astrometry   ____ Meteor Studies

List Regular Observation Site(s). Longitude, latitude, and altitude may be determined using topographic maps.

Name	Longitude	Latitude	Altitude
1. __________________________	________	________	________
2. __________________________	________	________	________
3. __________________________	________	________	________
4. __________________________	________	________	________

Provide the information requested on telescopes you expect to use including the units of measurement. Indicate the site numbers (from the list above) where the telescope has a permanent mount or where a portable mount is regularly used for visual (V), photographic (PG), and/or photoelectric (PE) observing. Binoculars users should state the power and aperture (e.g., 7x50) with the word binoculars under telescope type and skip the next two columns. Meteor observers should write meteor and visual, photographic, or radio under telescope type and give the site numbers where these observations are usually made.

Telescope Type	Aperture	Focal Length	Mounting Site # Perm. Port.	Observing V PG PE
__________________	________	________	________ ________	__ __ __
__________________	________	________	________ ________	__ __ __
__________________	________	________	________ ________	__ __ __
__________________	________	________	________ ________	__ __ __

List equipment planned for use in photography <u>not already listed</u> as a telescope.
This can include Schmidt or aerial cameras or interchangeable lenses belonging
to your photographic system.

Camera   Focal Length   f/ratio   Notes

__________   __________   ______   ________________________________

__________   __________   ______   ________________________________

__________   __________   ______   ________________________________

__________   __________   ______   ________________________________

<u>Photometric Equipment:</u>

Photomultiplier Tube _______________________   Cooled ____   Uncooled ____

Electronics: Photon Counting ___________   Analog ________

<u>Miscellaneous Accessories:</u>

Diffraction Grating Source or Manufacturer   ______________________________

________ gr/mm, blaze order _______

Prism: Glass Type ______   Apex Angle _________

Rotating Meteor Shutter Chop Rate ___________

I understand that the data I contribute to the International Halley Watch may
be used by IHW Archive users and that my contribution will be acknowledged in
the usual manner in any publications resulting from such use. I understand
that I may also publish my data in any manner I choose.

____
Signature   Date

                          Novice   Moderate   Expert
Level of Observational Experience
General Astronomical Observations   ____   ________   ______
Comet Observations
Meteor Observations

Are you planning on traveling to the southern hemisphere to observe Halley's
Comet in March or April 1986? Yes ___   No___

Additional Comments:

To

Nicholas T. Bobrovnikoff

His pioneering, comprehensive studies of
Comet Halley after its 1910 perihelion
passage have laid the groundwork for
research during the current apparition.

ABSTRACT

This manual describes the International Halley Watch, comets and observing techniques, and provides information on periodic Comet Halley's apparition for its 1986 perihelion passage. Part I gives detailed instructions for observation projects valuable to the International Halley Watch in six areas of study: (1) visual observations, (2) photography, (3) astrometry, (4) spectroscopic observations, (5) photoelectric photometry, and (6) meteor observations. Part II includes an ephemeris for Comet Halley for the period 1985-1987 and star charts showing its position from November 1985 through May 1986.

FOREWORD

This manual has been written for the advanced amateur astronomer. Part I provides instructions on the proper methods of generating meaningful scientific data on comets. The novice can learn general observing techniques while learning the methods in this manual. Part II contains an ephemeris and star charts for finding the comet and making observations of it.

This manual does not teach basic observing, telescope adjustment, or data reduction techniques. There are many books available for such purposes. It should be stressed that the most important single thing an amateur can do to advance his skills is to use them. Practice provides part of the training necessary to advance in the techniques of skillful, scientific observing. Both novice and experienced observers should observe comets as they appear in preparation for Comet Halley's apparition in 1985-86.

It is a sad fact that many professional astronomers are unaware of the careful, professional-quality work which amateur astronomers are capable of and have, in some cases, been doing for years. There is now a movement to call people who practice astronomy without pay "nonprofessionals" in an effort to improve the image of the hardworking, dedicated amateur who does reputable research. While "amateur" and "nonprofessionals" are both accurate, the latter is less fluent and has not been used in the text. It is my hope that activities like the IHW Amateur Observation Net will demonstrate to professionals that their unpaid fellow astronomers--amateurs--are worthy of the respect sought with the "nonprofessional" noun.

Amateurs have made and can continue to make important contributions to cometary research. Dedication to the effort is all that's required.

S. E.

ACKNOWLEDGEMENTS

Many people have had a hand in the creation of this manual. Ray Newburn, Leader of the International Halley Watch (IHW) encouraged its development and critiqued it in early drafts. Zdenek Sekanina and Mo Geller gave valuable advice. Charles Morris, John Bortle, Michael Hendrie, and Dave Meisel supplied useful input which led to great improvements in the text. The IHW professional Discipline Specialists made sure the sections describing observations useful to them were satisfactory. Tom Greska and Rick Shaffer were the guinea pigs when the time came for prospective users to read the manual. Wenonah Wells, Darlene Phillips, and Jackie Green put up with many changes, additions, and corrections in typing early versions of the manuscript, and their hard work is appreciated. Thanks are due them all for their help.

Special thanks are due to the American Association of Variable Star Observers and the British Astronomical Association for permission to reproduce portions of their excellent star atlases. Complete copies of these atlases are available from these associations; addresses will be found in Appendix I. I am grateful to Sky and Telescope magazine, Lick Observatory, and the U. S. Naval Observatory for permission to reproduce several illustrations and to Griffith Observer and the Royal Astronomical Society of Canada Observer's Handbook for permission to reproduce data tables.

IHW AMATEUR OBSERVER'S MANUAL

Part I - Methods

CONTENTS

Contents page for Part II is in the second half of this volume,
following page B-5.

TABLES

FIGURES

1. INTRODUCTION

Halley's Comet has a long and colorful history. The earliest definite record of its appearance is by the Chinese in 240 B.C. With the exception of the apparition in 164 B.C., written records of its passage have been found for every apparition since then (Table 1-1).

Though comets have long been feared as harbingers of catastrophe-- Jerusalem fell four years after Halley's return in 66 A.D., and World War I started four years after Comet Halley's last perihelion passage in 1910-- people often ignore the fact that one man's disaster is another man's good fortune. For example, the fall of King Harold of England to the Normans in 1066 AD with the appearance of Halley's Comet was a very happy event to the Normans, though the English didn't like it. The American public will likely debate for years whether the resignation of President Richard Nixon after the 1973-1974 appearance of Comet Kohoutek was good or bad, and for whom. Of course, many more great events in human history have occurred with no bright comet present than with one visible. Educated people no longer believe in comets as precursors of human events.

The scientific study of comets started with Tycho Brahe's study of the Great Comet of 1577 (not Halley's). Tycho showed that the comet moved through space at a distance greater than the Moon's and was not a "fiery exhalation" in the atmosphere as Aristotle and the scientists of earlier times believed. Johannes Kepler believed Halley's Comet moved in a straight line orbit when he computed its motion after its 1607 apparition. Isaac Newton later showed that the comet of 1680 moved in a nearly parabolic orbit. Using observations of many past comets and Newton's methods, Edmond Halley (rhymes with alley) computed the orbits of twenty-four comets. He noticed groups of several comets that had similar apparent orbits. One of the three orbital groups he noticed had comet appearances in 1531, 1607, and 1682. The comet of 1456 may also have been a member of this group. This evidence led him to conclude that these were appearances of the same comet moving in an elliptical orbit and to predict the comet's reappearance late in 1758 or early early 1759. Sixteen years after his death, Halley's predicted comet was recovered on Christmas, 1758, by amateur astronomer Johann Palitzsch, a farmer near Dresden. The comet was independently recovered a month later by professional astronomer Charles Messier. It has been known as Halley's Comet since then.

For the next century the study of comets turned to the glory of discovering them and computing orbits. The first comet photograph was obtained of Donati's Comet in 1858, but high-quality photos were not obtained until 1881 with Tebbutt's Comet and with the Great Comet of 1882 in the next year. The first spectrograms of a comet also were those of Tebbutt's Comet, obtained by William Huggins. Photographic photometry was even attempted by J. Janssen on this comet.

The stage was now set for detailed scientific studies of Halley's Comet during the 1909-11 apparition. Halley was recovered photographically on September 11, 1909, by Max Wolf at Heidelberg. Two months later, acting on E. E. Barnard's suggestion of the 1890s, the Comet Committee of the

TABLE 1-1

Halley's Comet Perihelion Dates

Year	T(E.T.)
2061	2061 Jul. 29
1986	1986 Feb. 9.6613
1910 II	1910 Apr. 20.17771
1835 III	1835 Nov. 16.43871
1759 I	1759 Mar. 13.06075
1682	1682 Sep. 15.28069
1607	1607 Oct. 27.54063
1531	1531 Aug. 26.23846
1456	1456 Jun. 9.63257
1378	1378 Nov. 10.68724
1301	1301 Oct. 25.58194
1222	1222 Sep. 28.82294
1145	1145 Apr. 18.56090
1066	1066 Mar. 20.93405
989	989 Sep. 5.68757
912	912 Jul. 18.67429
837	837 Feb. 28.27000
760	760 May 20.67126
684	684 Oct. 2.76682
607	607 Mar. 15.47581
530	530 Sep. 27.12998
451	451 Jun. 28.24911
374	374 Feb. 16.34230
295	295 Apr. 20.39842
218	218 May. 17.72347
141	141 Mar. 22.43405
66	66 Jan. 25.96014
12 B.C.	- 11 Oct. 10.84852
87 B.C.	- 86 Aug. 6.46171
164 B.C.	-163 Nov. 12.56604
240 B.C.	-239 May 25.11796
315 B.C.	-314 Sep. 8.52367
391 B.C.	-390 Sep. 14.36897
466 B.C.	-465 Jul. 18.23879
540 B.C.	-539 May 10.82702
616 B.C.	-615 Jul. 28.50346
690 B.C.	-689 Jan. 22.27922

This table is adapted from Yeomans and Kiang (1981) and Yeomans (1977).
Perihelion passage times (T) are in Ephemeris Time and the Julian calendar
is used for dates earlier than 1607.

Astronomical and Astrophysical Society of America proposed worldwide, coordinated observations of Halley's Comet. The organization was swamped with data even though some observatories refused to cooperate. Lack of funds and manpower prevented proper use of the large amount of data submitted. Comprehensive studies of some parts of the data were finally published in 1931 by Nicholas T. Bobrovnikoff of the Lick Observatory and, in 1934, by C. D. Perrine of the Cordoba Observatory.

The 1910 apparition was not without the drama only human nature can provide. The predicted passage of the Earth through the comet's tail prompted fears of death by poison gas and the end of the world. (Earth just missed the dust tail and missed or passed only through the fringes of the ion tail.) There was business in sales of gas masks and "comet pills." That sheriff's deputies had to stop the sacrifice of a virgin in Oklahoma is not a true story.

It is natural to expect human nature to provide many shenanigans at the next apparition (and last seen as recently as 1973 with Comet Kohoutek). We can also expect a much better informed, more sophisticated, curious public anxious for factual information.

The International Halley Watch has been organized to coordinate observations and archive data and to provide factual information to amateur astronomers, the news media, and the public.

The operational goals of the IHW are to standardize observational techniques where this is useful, to promote simultaneous observations of Halley's Comet by many techniques, and to organize closely spaced time-sequenced observations during the apparition. The IHW will work closely with representatives of the science teams operating spacecraft flying by Halley, observing it from earth orbit, and observing with airborne or rocket-borne instruments.

The ground-based effort will be subdivided among seven professional disciplines. Professional astronomers will be organized by seven Discipline Specialist Teams (Table 1-2) and their staffs:

(1) Large Scale Phenomena Studies will use wide-angle photography to study the dust tail and to examine the interaction of the solar wind and the ion tail.

(2) Near-Nucleus Studies, by using photographic and electronic imaging of structures in the coma, are expected to yield data on the nucleus (rotation rate, active regions, surface structure, etc.), the inner coma (interactions of dust and gas with solar radiation), and the general activity of the comet.

(3) Spectroscopy and Spectrophotometry will generate data on the composition and physical state of the coma and tail which will help efforts to model the nucleus.

(4) Photometry and Polarimetry are expected to determine the abundance of major volatile and nonvolatile components of the comet and the physical mechanisms acting on them to make the comet behave as it does.

(5) <u>Radio Science</u> will provide further data on the physical processes
and composition of the comet through the study of cometary chemical
species observable at radio wavelengths.

(6) <u>Infrared Spectroscopy and Radiometry</u> generate additional information on the composition and physical state of the coma and allow
quantitative determination of the amount and spectral distribution
of the comet's thermal radiation, giving information on the temperature, composition, and size of particles released by the comet.
Some data may also be obtained on gaseous components of the coma.

(7) <u>Astrometry</u> will provide precise positional observations required
for orbit and ephemeris computations, dynamical modeling of the
nucleus to explain observed nongravitational effects on the comet's
motion, and "nucleus" diameter estimates.

In addition to these professional observation nets, amateur astronomers
who wish to contribute will be organized to make observations of Halley's
Comet. Only they will make observations directly comparable to those made
at the last apparition. This organization is described in the next section.

Very complete models of Halley's Comet can be expected after all the
data are analyzed. Undoubtedly many new and surprising discoveries will be
made as the result of this first attempt to study a complete cometary
apparition by every modern technique available. With the IHW Amateur
Observation Net, amateur astronomers can expect to make contributions to
this effort.

TABLE 1-2

Discipline Specialist Team Members

<u>Discipline</u>	<u>Discipline Specialist Team</u>
Large Scale Phenomena	J. C. Brandt, M. B. Niedner, J. Rahe
Near-Nucleus Studies	S. Larson, Z. Sekanina, J. Rahe
Spectroscopy and Spectrophotometry	S. Wyckoff, P. Wehinger, M. C. Festou
Photometry and Polarimetry	M. F. A'Hearn, V. Vanysek
Infrared Spectroscopy and Radiometry	R. F. Knacke, T. Encrenaz
Radio Science	W. M. Irvine, F. P. Schloerb, E. Gerard, R. D. Brown, P. Godfrey
Astrometry	D. K. Yeomans, R. M. West, R. S. Harrington, B. Marsden

2. THE AMATEUR OBSERVATION NETWORK

From the very beginning, organizers of the International Halley Watch recognized that amateur astronomers could make valuable contributions supplementing the comprehensive professional observations being planned. Because of the large number of amateurs, the interference of weather with observations would be minimized and geographic longitude coverage would be more complete than for the smaller number of professionals participating. Also, amateurs are not constrained by telescope time allotments or other duties which might limit a professional astronomer's time. Finally, there are some observations of Halley's Comet and related phenomena which are simply more easily done by amateurs, and more comprehensive coverage is possible with their help.

With this justification, the IHW was organized to include a Coordinator for Amateur Observations (CAO) whose job is to coordinate the activities of the amateur observation net. In addition, he is to provide information and instructions to amateur contributors so that their observations make useable and valuable additions to the IHW data set.

In order to keep amateur contributions manageable, an intermediate level organization will be established to spread the workload. In the United States, individual amateurs will submit their observations to a Recorder who will judge the quality and completeness of the material and then submit, on a regular basis, a collection of various observers' data to the IHW. Recorders are also responsible for answering questions on observational technique. Each observer should retain a copy of the report submitted to the Recorder in case clarification or duplication is necessary. Recorders' names and addresses will be published in an issue of the IHW Amateur Observer's Bulletin. Introductory issues are available from the IHW.

The CAO will work with the astronomical organizations in other countries to establish methods for handling data acquired by their nationals.

All the data collected by amateurs world-wide will be examined en masse at one or two meetings in 1986. The CAO, Recorders, staff of the International Comet Quarterly, and leaders of the real-time observation net* will participate. The data will then be dispersed to the IHW archives and/or the Discipline Specialists.

Several areas of study have been identified to which amateurs can make significant contributions. These include visual observations, photography, astrometry, spectroscopy, photoelectric photometry, and meteor studies. Detailed descriptions and observational methods are given in separate sections elsewhere in this manual.

Attempts will be made to have meetings of contributors to the amateur

---

* A Real-Time Observation Network is being established to provide the IHW with current data on the appearance of the comet. It is being operated much like the professional nets.

net at regularly scheduled regional gatherings of amateur astronomers. In addition, contributors will be kept informed of current events with the _Bulletin_. An early issue will carry more details on plans for handling the data.

Halley's Comet will first appear in amateur telescopes in mid-1985. However, all observers are encouraged to practice their technique on any comets which appear before that time, and to participate in the scheduled trial-run in 1984.

Problems of any type or questions on amateur net operations should be submitted to the CAO or to the appropriate Recorder.

3. COMETARY ASTRONOMY

Cometary Phenomena

In the centuries that comets have been observed, surprisingly little
has been learned about the details of their origin, evolution, and the
processes occurring in them. Most of our detailed knowledge of comets has
been acquired in the past few decades, and even this is patchy. One goal of
the International Halley Watch is to collect and archive the most complete
set of cometary data ever acquired on a single comet. Halley's Comet is an
excellent target for this activity because, among all the periodic comets
(those with well-known periods less than 200 years long) with predictable
orbits, it alone exhibits virtually all the phenomena seen in other periodic
and long-period (periods greater than 200 years) comets.

Observationally, a comet can be separated into three components: the
nuclear region, the coma, and the tail. Meteors and the zodiacal light are
generally agreed to be related to comets.

The nucleus is the source of all cometary phenomena. This tiny member
of the solar system, almost never directly observed, generates some of the
largest phenomena (comet tails) and some of the smallest objects (dust
particles and free molecules and atoms) observed in the solar system.

The nucleus is believed to be a fluffy snowball with dust mixed in.
The snow is not pure water ice but, rather, a mixture of frozen gases that
includes carbon dioxide (CO_2), hydrogen cyanide (HCN) and others containing
carbon and sulfur in addition. Some of these molecules are believed to be
mixed with or trapped within the water ice and dust. This picture has
developed on the basis of spectrocopic studies of the molecules and the
continuous spectrum of the tail.

The proportions of gas and dust in the nucleus are not well-known but
appear to vary considerably from comet to comet. The nucleus diameter is
believed to range from a few hundred meters to 10 km in diameter. The
density of the nucleus is believed to be approximately that of water: one
gram/cubic centimeter. The occasional observations of the fragmentation of
a cometary nucleus suggest that it has little internal strength.

During its passage through the inner solar system, the heat of the sun
causes the ices to sublimate (change from solid to gas, directly, without
changing to a liquid first). Halley's nucleus loses material at a rate per
orbital revolution that, if spread all over the nucleus, would form a shell
of material about one meter thick. In fact, it appears that portions of the
nucleus are quiescent while "hot spots" are the primary source of material
in the coma. Nongravitational forces which affect the comet's orbital
motion are a result of the sublimation process.

The coma is an approximately spherical halo of material surrounding
the nucleus. Gas streaming from nuclear hot spots as the ices sublimate
carries dust particles with it into this tenuous cometary atmosphere.

A coma does not generally form until the nucleus is within three
astronomical units (AU; 1 AU $\approx$ 149,600,000 km $\approx$ 93,000,000 miles) of the

sun. Faint comets usually generate smooth-appearing comas. More active
comets like Halley's often show jets or fountains emanating from the central
condensation (innermost, brightest portion of the coma) surrounding the (invisible) nucleus. Envelopes or hoods are often seen concentrically placed on
the central condensation. Envelopes are being used to find the direction of
the rotation axis of Halley's Comet and its rotation period of perhaps ten
hours. This research is based on drawings of exceptional quality made during
the 1835 apparition and on digitally processed photographs taken in 1910.

     The coma and nucleus together are referred to as the head of the comet.
Surrounding the head is a huge cloud of atomic hydrogen gas emitting ultraviolet light. This hydrogen envelope can be one to ten million kilometers
in size.

     The sun affects a comet in more ways than simply supplying heat to
sublimate the nuclear ices. Electromagnetic radiation (radio, infrared,
visible, ultravioltet, x-ray, etc.) from the sun can interact with material
released by the comet by affecting its electric charge and internal energy
and by acting as a force to affect its motion after leaving the head. The
solar wind and magnetic fields it carries play a role in shaping the tail.

     A comet's tail is observed to have two components. The ion tail con-
sists of molecules released by the nucleus that have been ionized (forced to
lose an electron and, thus, acquire a positive charge) by solar ultraviolet
and x-radiation.

     The solar wind, a stream of electrically charged particles (ions
[charged atoms] and electrons) blowing out from the sun at several hundred
kilometers per second, carries magnetic fields which drag cometary ions
along with them away from the sun. Solar electromagnetic radiation continuously excites the molecular ions causing them to glow in characteristic
wavelengths. The ion tail has an emission spectrum with bright lines
primarily at the blue end of the spectrum. This is the cause of its bluish
appearance in color photographs.

     On a biweekly or weekly basis, the ion tail of a comet may be discon-
nected from the head of the comet. This disconnection event appears
to occur because the polarity of the magnetic field in the solar wind
changes, and the postulated weak cometary magnetic field reacts to this
change. A new ion tail can be rebuilt in as little as 30 minutes after the
disconnection.

     The dust tail is generated by a different process. Dust about one
micrometer in size and, in part, silicate in composition is carried into the
coma by the "wind" of molecules released as the nucleus sublimates. Solar
radiation pressure affects dust particles the way wind drives a sailboat.
The dust is pushed away from the sun while at the same time moving with the
comet's orbital motion. The result is a curved dust tail. Dust is spread
in the plane of the comet's orbit outside the orbital path.

     When the comet is observed from out of the plane of its orbit, the dust
and ion tails are well-separated because ions respond strongly to the high-
velocity, almost radial solar wind. The low inclination of Halley's orbit
will make seeing separate dust and ion tail components difficult. For a few

days around the time that the Earth crosses a comet's orbital plane, projection effects may allow observers to see an antitail which appears to point in the direction of the sun. The antitail is due to tail dust in the orbital plane well behind the comet, but which appears opposite the tail because of projection effects.

Because sunlight illuminates the dust tail, a solar spectrum is scattered back to observers across its broad sweep. This tail may appear distinctly reddish in binoculars.

Banding or streaks photographed in the dust tail may be synchrones, groups of particles released at the same time, or striae, formed from parent particles released at the same time which later disintegrate and spread out in space. The leading edge of the tail is usually close to a syndyname where all the particles respond to equal force.

Long after a periodic comet has left the inner solar system, its effects can still be observed. Particles released by the comet during its return to perihelion are affected by radiation pressure and the gravitational fields of the planets. Most micrometer-sized and smaller particles are blown out of the solar system. Eventually, the submillimeter and larger particles may be perturbed into orbits that intersect the Earth's. When the Earth is in the vicinity of the intersection, "falling stars" known as meteors are seen as the particles (called meteoroids when in space) burn up in the Earth's atmosphere. Usually they are first seen at heights of 80 to 100 km and disappear at a height of about 50 km. Particles reaching the ground are called meteorites. However, cometary meteoroids are generally fluffy and burn up in the Earth's atmosphere before reaching the ground. Most meteorites that have been recovered probably originated in the asteroid belt between the orbits of Mars and Jupiter.

Halley's Comet is believed to be the parent body of the η Aquarid meteor shower in May and the Orionid meteor shower in October. The meteors appear to radiate from Aquarius and Orion, respectively, because of the orbital geometries at the times they collide with the Earth's atmosphere.

Particles from the comet's tail slowly spread throughout the solar system and scatter sunlight. (Scattered light is spread in all directions, sometimes more strongly in certain preferred directions depending on particle size and shape.) Multitudes of particles in the one micrometer to $1/10$ millimeter size range may be the source of the triangular glow on the sunrise or sunset horizons known as the zodiacal light pyramid (Table 3-1). The zodiacal light is faintly visible over the entire sky, and the pyramids which are its brightest parts can be seen best when the sun is more than 18° below the horizon after sunset and before sunrise. (When the Sun is 18° below the horizon, no part of the atmosphere seen by an observer is illuminated by sunlight.) Opposite the sun in the sky, a large, faint glow known as the counterglow or gegenschein can sometimes be seen when the sun is well below the horizon (Table 3-2). The gegenschein is fainter than the zodiacal light pyramids but is brighter than the very faint zodiacal band which connects the pyramids to the gegenschein. All aspects of the zodiacal light are placed nearly symmetrically with respect to the ecliptic.

More detailed explanations of these phenomena can be found in many of the references in the bibliography or in textbooks on astronomy.

TABLE 3-1

Zodiacal Light Pyramid Visibility

<u>Season</u>*	<u>Time</u>
Winter	Morning and Evening
Spring	Evening
Summer	Evening and Morning
Autumn	Morning

* Northern hemisphere given. Optimum times for the southern hemisphere are opposite those in the north. The pyramid is easily visible morning and evening all year round at the equator. In the temperate zones, the inclination of the ecliptic affects the ease of observation of the pyramid. It is less optimally placed for observation in northern hemisphere spring mornings and autumn evenings.

TABLE 3-2

Gegenschein Visibility

<u>Month</u>	<u>Constellation</u>	<u>Comments</u>
January	Gemini	Just East of the Milky Way
February	Leo	Near Regulus
March	Leo-Virgo	
April	Virgo	Near Spica
May	Libra	
June	Scorpius	In the Milky Way
July	Sagittarius	Just East of the Milky Way
August	Capricornus	
September	Aquarius-Pisces	
October	Pisces	
November	Aries-Taurus	Southwest of Pleiades
December	Gemini-Taurus	In the Milky Way

This table is adapted from <u>Griffith Observer</u>, Paul Roques and Patricia Whitt (1971).

Glossary

<u>Antitail</u>: Projection effects, when the Earth crosses the orbital plane of the comet, sometimes make a portion of the comet's tail appear to point towards the sun.

<u>Apparition</u>: The period of time that a celestial object is visible from Earth.

<u>Coma</u>: The volume containing gas and dust around the nucleus of the comet which has not yet been swept into the tails by the solar wind and solar radiation pressure.

<u>Dust Tail</u>: Solid dust particles (blown off the nucleus of the comet as it sublimates), responding to solar radiation pressure and their orbital motion, are pushed away from the nucleus. The dust tail is seen because of sunlight scattered by the dust particles.

<u>Ecliptic</u>: The path of the sun in the sky projected on background stars.

<u>Fluorescence</u>: The emission of light of a longer wavelength after absorption of shorter wavelength electromagnetic radiation by atoms, molecules, or ions.

<u>Gegenschein</u>: Literally meaning "counterglow," this phenomenon of the zodiacal light is due to sunlight back-scattered from interplanetary dust located outside the Earth's orbit and opposite the sun in the sky.

<u>Head</u>: The nucleus and coma of the comet are collectively referred to as the head.

<u>Hydrogen Envelope</u>: Seen only in ultraviolet light, this gigantic cloud of atomic hydrogen surrounds the comet's head.

<u>Ion, Ionize</u>: An ion is a neutral atom or molecule which acquires additional positive or negative charge. Solar ultraviolet radiation is the principal reason neutrals become ionized in comets.

<u>Ion Tail</u>: The parent molecules released by the nucleus are ionized by sunlight and dragged away by the magnetic field carried by the solar wind to form the ion tail. The tail is seen by the light of fluorescing ions.

<u>Meteor</u>: The rapidly moving streak of light caused by a particle as it burns up in the Earth's atmosphere.

<u>Meteorite</u>: A natural particle reaching the surface of the Earth from space after traveling through the Earth's atmosphere.

<u>Meteoroid</u>: A natural particle in space before it enters the Earth's atmosphere.

<u>Nongravitational Forces</u>: Forces changing a cometary orbit that are not due to gravitational effects; usually identified with rocket-like forces on the nucleus (the so-called "rocket effect").

<u>Nucleus</u>: The source of all cometary phenomena, the nucleus is believed to be a snow ball of frozen gases and dust.

<u>Parent Molecules</u>: Water (H_2O), carbon dioxide (CO_2), hydrogen cyanide (HCN), and other molecules containing carbon and sulfur are believed to be the source molecules for many of the neutral and ionized atomic and molecular species observed in the coma and tail of a comet.

<u>Perihelion</u>: The point in an orbit around the sun which is closest to the sun.

<u>Perturbation</u>: Gravitational effects on the orbital motion of an object by masses other than the sun (usually major planets).

<u>Plasma</u>: A "gas" of positive and negative ions.

<u>Plasma Tail</u>: A different name for an ion tail.

<u>Radiation Pressure</u>: Electromagnetic radiation (e.g., light, infrared, x-rays, radio, ultraviolet, etc.) has the property of being able to transfer momentum - push - materials away from the source of the radiation.

<u>Scattering</u>: Small particles (one micrometer to $1/10$ millimeter in size) have the property of not simply reflecting light and making shadows but actually scatter the light that illuminates them in all directions. In some situations forward-scattered light, appearing where a shadow would be expected is actually brighter than back-scattered ("reflected") light.

<u>Solar Wind</u>: Ionized gases carrying magnetic fields are blown off the sun at speeds in the range of 450 km/sec.

<u>Striae</u>: Narrow, rectilinear structures sometimes seen in the dust tail. They are made of particles that were released at the same time from the nucleus and later disintegrate into fragments.

<u>Sublimate</u>: The change of state directly from solid to gas without going through a liquid phase.

<u>Synchrones</u>: The loci of particles released from the nucleus simultaneously. They are sometimes seen in the dust tail as straight or moderately curved structures.

<u>Syndynames</u>: The loci of particles in the dust tail that are subjected to equal force.

<u>Tail</u>: A general term used to describe the ejecta (ions and dust) streaming out from the comet head opposite the sun.

<u>Zodiacal Band</u>: The faint glow seen along the ecliptic connecting the zodiacal light pyramids to the gegenschein.

<u>Zodiacal Light</u>: A general glow throughout the sky caused by sunlight
scattered by interplanetary dust. It is brightest near the sun and along
the ecliptic. The zodiacal light pyramids are often referred to as the
zodiacal light.

<u>Zodiacal Light Pyramid</u>: This triangular glow seen on the western horizon
after evening twilight and on the eastern horizon before morning twilight
is the brightest component of the zodiacal light.

4. INTRODUCTION TO COMET OBSERVATIONS

<u>Scientific Observing</u>

One purpose of the International Halley Watch is the collection and archiving of data on Halley's Comet at this apparition. To be useful, these data must be delivered with a sufficient amount of background information to allow physical interpretation. Insufficient calibration data render the comet data useless or nearly useless. It cannot be emphasized enough that acquiring data lacking necessary calibration is a waste of valuable time and energy.

Amateur observation Recorders will examine all the observational data and background calibration submitted to them to determine their quality and suitability for inclusion in the IHW archives. The data will be passed on to the Coordinator for Amateur Observations for distribution to the concerned professional Discipline Specialist and/or inclusion in the archives.

It is strongly recommended that all participants maintain a bound log-book dedicated specifically to observations of Halley's Comet. This logbook can serve as a valuable permanent record not only of the observations them-selves but also of the necessary background observations and personal impressions - scientific and emotional - of the apparition.

Report forms and a glossary explaining the information requested for the various observations described will be found at the end of their respec-tive sections in Part I and in the first section of Part II. Observers should reproduce the ones they intend to use (by Xerox, for example) and then use the reproductions for submitting observations. Space is included for standard calibration observations for each night's data set. Report forms can be filled out when an observation is made or the data can be transcribed later from the logbook as long as the calibration data taken with the cometary data are recorded and transferred together.

The techniques of astronomical investigation, whether they rely on the eye, camera, or electronics, must first be learned and then practiced regu-larly to maintain proficiency. Also, an ongoing series of synoptic observa-ations (that is, observations to provide a general view) is more valuable than scattered individual observations. It is best to do one project well, continuing one type of effort throughout the apparition, rather than attempt a variety of projects. A variety of projects may be well done (though usually none are done as well as an individual one which receives all of an observer's attention), but the group is often less valuable than concentra-tion in one area.

It is strongly recommended that Halley Watch contributors start prac-ticing their comet observing techniques immediately. Through the year there are usually several comets available for observation. These provide excel-lent targets of opportunity to practice observing techniques, from making drawings at the telescope to darkroom procedures to computation of results. It is better to learn what mistakes are possible in advance of the main event rather than find out at an inopportune moment during the event. Practicing early will also provide valuable experience that will make for higher quality observations when Halley's Comet is visible.

Besides submitting data to the IHW during the scheduled trial-run in
1984, data taken on other comets can be submitted to the International
Comet Quarterly, the Comet Section and/or Journal of the Association of
Lunar and Planetary Observers, Sky and Telescope magazine, J. Bortle of
the W. R. Brooks Observatory, and to numerous other national and international organizations. (A list of addresses will be found in the Appendix.)

Dark Adaptation, Averted Vision, and Eye Sensitivity

Many amateur astronomers don't realize the difference full dark adaptation makes in viewing the sky and faint objects. While twenty to thirty
minutes are necessary for initial dark adaptation, significant increases in
the eyes' sensitivity occur with extended stays in the dark; one to two
additional hours produce a noticeable increase in the discernability of weak
sources when observed from dark-sky sites.

The long periods required for full dark adaptation do not mean one must
sit around in a closet doing nothing. It does mean that trips into illuminated areas must be abandoned, and the use of bright red flashlights should
be curtailed.

Celestial observations and the use of muted red lights for note taking
or map reading are fine uses of time.

Low-light sensitivity is enhanced by the avoidance of strong sun and
fluorescent light during daylight hours. A good pair of sunglasses will
serve to cut the strength of light outdoors. Military surplus or fluoro-
scopic red goggles serve very well in starting the dark adaptation process
(even in daylight) and in maintaining adaptation after dark when visits to
lighted areas are unavoidable. Use of a dark hood during observations may
also prove helpful.

Formal Halley Watch observations of the comet and atmospheric transparency should not start until at least initial dark adaptation has occurred.
The IHW recognizes that this goal cannot always be reached, but observers
are strongly encouraged to make every effort to attain it.

Averted or indirect vision is useful when trying to see an object at
the limit of the eyes' sensitivity. Because the high resolution, color
sensitive cones of the retina are situated on the optical axis of the eye,
it is possible to see fainter sources by looking 10° - 20° away from the
source, while holding attention on the source. This allows light to fall
on the much more light sensitive (but color-insensitive) rods which are
found in greater concentration off-axis.

Averted vision should not be used for visual magnitude and coma diameter
estimates of the comet because it is difficult to repeat positioning of the
comet and comparison stars on the same area of the retina. Data for other
projects, especially in making drawings, may be improved by the use of
averted vision.

Atmospheric Transparency and Sky Brightness

Many Halley Watch observations are sensitive to the transparency of the
Earth's atmosphere and to background sky brightness due to artificial and
natural sources. Observations of the comet's tail, the size and magnitude
of its coma, photographic exposures, photoelectric measurements, and meteor
visibility are among those that are strongly influenced by transparency.

Because both urban sites with bright skies and rural sites with dark
skies will be used for various observations, a method to standardize
estimates of atmospheric transparency and sky brightness is necessary.
Such a method is also useful when the Moon is up during comet observations.

The "Faintest Star" column included on some observation report forms
is for reporting the magnitude of the faintest star visible to the naked
eye on the chart showing the comet's position for the day observed. The
magnitude reported should be based on a star at a similar altitude above
the horizon as the comet and as close to it in the sky as practicable. For
meteor observers, the magnitude of the faintest star visible in the center
of their field of view should be given.

Large-Scale Sky Measurements

Estimates of angular distances in the sky are notoriously unreliable.
For the purposes of the International Halley Watch, reliable, quantitative,
visual measurements of comet tail length are required.

Eq. (1) on page 5-4 is the most reliable method of determining angular
distances in the sky. Other methods given in the Visual Observations section
may also be used.

Participating observers may wish to construct a simple Sky Crossbow
("Projects for May with a Sky Crossbow," Sky and Telescope, May 1981,
p. 417) requiring only a yard or meterstick, rod, and string. Attach the
middle of a flexible yardstick or meterstick to the end of a rod so that
your eye is 57 inches or 57 cm, respectively, from the middle of the stick.
Wood dowel or PVC pipe are good materials for this purpose. The rod can be
made collapsible for easier storage.

Attach the string to both ends of the stick so that the stick bows
towards the eye-end of the rod with 2 $^3/_4$ inches or 2.75 cm, respectively,
from the string to the middle of the stick (Fig. 4-1). Luminous paint can
be added to the inch or centimeter marks if desired. All one has to do is
aim the Crossbow at the object of interest and determine the number of inches
or centimeters (which equals the number of degrees) across the object. A red
flashlight may help in reading the scale, but be careful to avoid ruining
dark adaptation.

Final calibration of the device can be made using the star separations
given in Table 4-1. Shorten the staff if required. Values in inches or
centimeters greater than given separations indicate that the staff is too
short.

A more compact but less accurate derivative of the crossbow can be
made with a string or light chain attached to the middle of a centimeter
ruler (Huling, 1981). The chain (preferred because it won't stretch) should
be about 57 centimeters long. The device is held so that the chain is
stretched taut from teeth or cheekbone to ruler. Final calibration should
be made on known star separations, reading the separation in centimeters as
degrees of arc. This device is less accurate than a sky crossbow because of
the greater eye focus compensation required to go from nearby ruler to sky
at infinity.

A string or chain stretched taut between outstretched arms can also be
used for large scale measurements if star separations are used for calibration. Beads or knots can be added at regular, known intervals to read the
angular distance off accurately. This method suffers from the disadvantage
that each observer must calibrate his/her own chain, and a large angular
distance spreads the observer's arms so the scale calibration varies with
hand separation and orientation of the arms with respect to the observer's
head. While this method is very convenient, it is not recommended.

Table 4-1

Sky Calibration Distances

Star Pair		Separation [°]
α Boo	α Vir	32.8
α Boo	β Leo	35.3
α Boo	ζ UMa	37.1
α Boo	α Lyr	59.1
α Lyr	α Cyg	23.8
α Aql	α Lyr	34.2
α Aql	α Cyg	38.0
α Aql	α Sco	60.3
α Ori	α CMa	27.1
α Ori	α CMi	26.0
α Ori	α Tau	21.4
α Tau	α Aur	30.7
α Cen	α Cru	15.6
α Cen	α Car	58.0
α Cen	α Eri	61.3

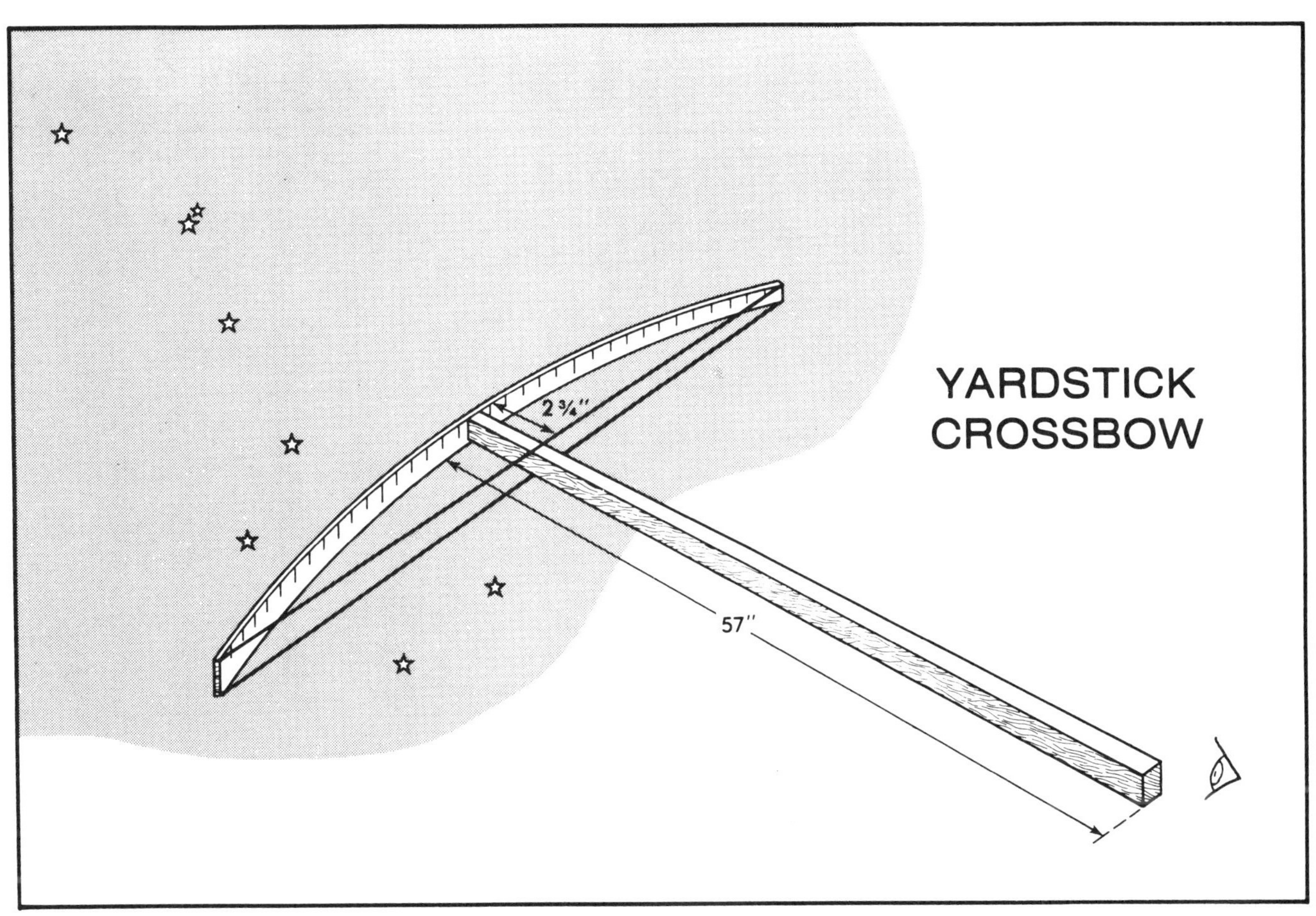

This simple device will allow angular distances to be measured directly on the sky. The only parts are a shaft 57 inchcs long and an ordinary yardstick bent slightly with a string as shown. Inches correspond to degrees. The eye-end of the long stick should be placed in contact with the observer's cheekbone; a flashlight will aid in reading the scale at night.

Figure 4-1. Sky Crossbow. Reproduced by Permission of <u>Sky and Telescope</u>

Universal Time

Universal time (UT), frequently called Greenwich Mean Time, is the local mean time at 0° longitude. It has been adopted by the International Halley Watch for use in designating the time of all observations and activities. Standardizing on this time system, broadcast on certain short wave radio frequencies by some national time services, will ease data recording and reduction problems. The accompanying map shows standard time zones around the world (Fig. 4-2).

To obtain UT from local standard time (LST; that is, clock time), first convert times after noon to a 24-hour clock by adding 12 hours to the clock time. Times before noon (LST) need no conversion. Then subtract the number of hours given in the table (with the map) from LST. The equations are:

$$\text{Before noon, UT = LST - (value from table)} \qquad (1a)$$

$$\text{After noon, UT = LST + 12 - (value from table)} \qquad (1b)$$

When UT exceeds 24 hours, subtract 24 and add one day to the date. Note that zones west of 0° longitude have a negative value in the table, so subtracting the negative value is equivalent to adding a positive value. In countries switching to daylight saving time (DST) or summer time, an additional one or two hours (depending on the country) must be subtracted from DST to get LST:

$$\text{LST = DST - (1 or 2)} \qquad (2)$$

Examples:

(1) For 10:35 pm in Western Australia (zone H)

   10:35 pm = 10:35 + 12 = 2235 LST

   Then UT = 2235 - 800 = 1435 UT

(2) For 5:14 am in Peru (zone R)

   5:14 am = 0514 LST

   Then UT = 0514 - (-5) = 0514 + 5 = 1014 UT

(3) For September 10, 12:48 am, daylight saving time in Alaska (zone W):

   12:48 am - 1 = 11:48 pm (standard time on September 9)

   = 11:48 + 12 = 2348 LST

   Now 2348 - (-10) = 2348 + 10 = 3348 UT

   To recover UT, subtract 24 and increase the LST date by one:

   UT = 3348 - 2400 = 0948 UT on September 10.

WORLD MAP OF TIME ZONES

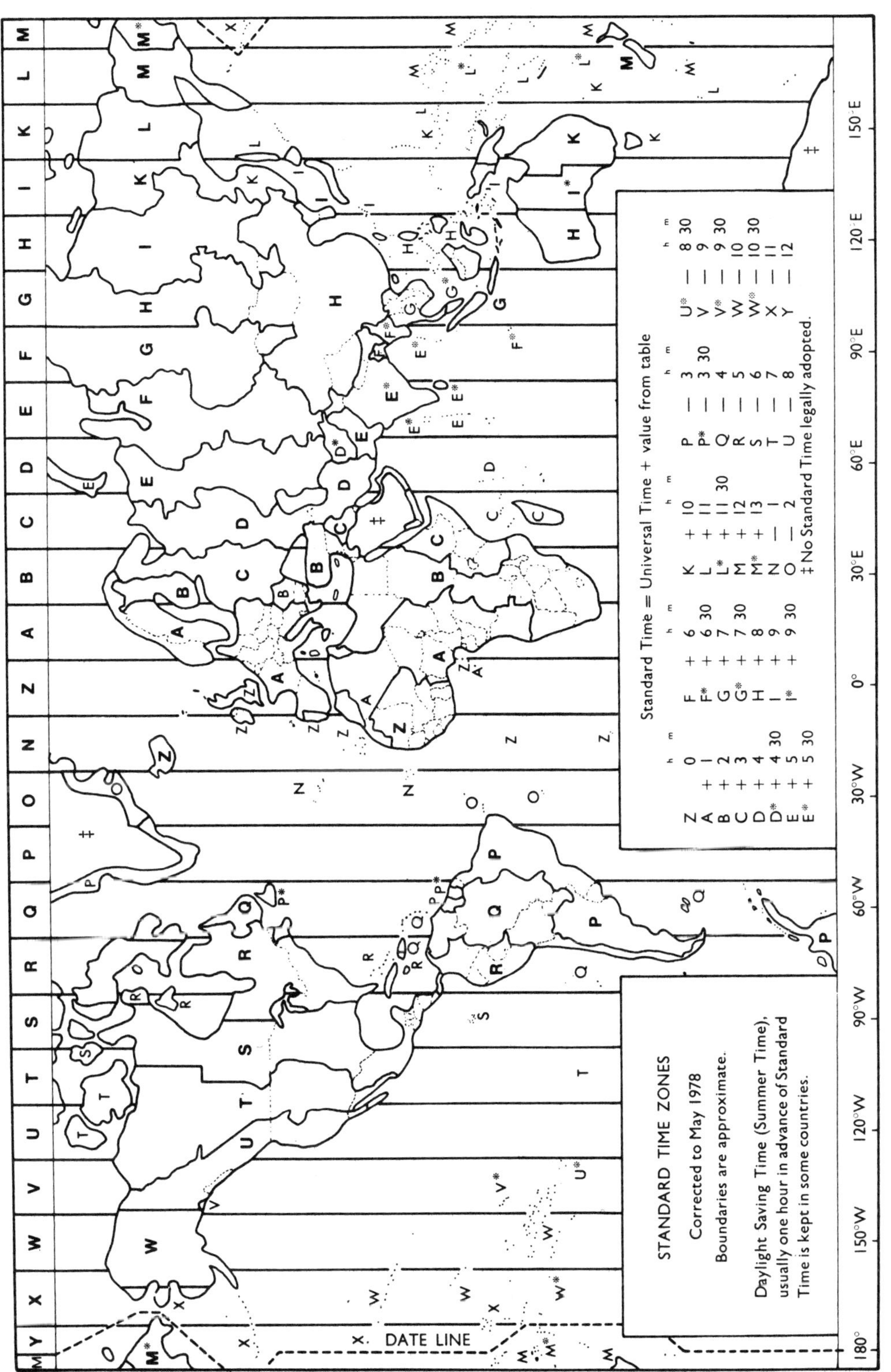

Figure 4-2. World Map of Time Zones, Taken from Astronomical Phenomena for the Year 1982 (Washington: U. S. Government Printing Office and London: Her Majesty's Stationery Office)

5. VISUAL OBSERVATIONS

Amateur astronomers contributing in the visual observation area provide
a very important service: their observations are the connection between the
current apparition and all previous apparitions of Halley's Comet. Except
for the photography done during the 1909-1911 apparition, all observations
of Halley's Comet have been visual: magnitude estimates, coma size, tail
studies, and drawings. Amateurs can continue these observations; few
professional astronomers will be making them.

The maximum scientific value of visual observations made by many different observers is attained when the many variables inherent in observational
techniques are minimized. The Halley Watch must insist that visual observations be made using the procedures given below. This standardization will
make interpretation of the data much easier. Also, the suggested methods
have been selected because they are the best available based on current
experience and analysis of comet observations.

Visual photometry of a comet can have three targets: the nucleus, the
head, and the tail of the comet. Photometry of the nucleus is difficult
because the true star-like nucleus is rarely seen, being invisible and/or
almost always confused with a false "photometric" nucleus or central condensation, and because it is seen against the bright background of the coma.
It is very difficult to do accurate visual photometry on the tenuous, low
contrast, filamentary tail of a comet. For this reason, tail photometry
is not part of the IHW visual program. Observers are, however, encouraged
to monitor the brightness of the central condensation and nucleus if visible
and to record the times of any abrupt changes in brightness. These magnitude
estimates and timings may prove very useful in the analyses of nucleus characteristics.

In the past, methods developed by Bobrovnikoff, Sidgwick, and Beyer
have been used to estimate cometary brightness. A study and comparison of
these methods led C. S. Morris (1979, 1980) to suggest a new method. All
these methods of visual photometry require the observer to memorize the
image of the comet and then move the telescope to comparison stars some
distance away. <u>Colored glass filters and so-called "nebular," "deep sky,"
"light pollution," "comet," or similar filters should not be used for
magnitude estimates.</u> The relation between visual magnitudes and filtered
magnitudes is not known, and scientifically interpreting filtered comet
magnitudes is not now possible nor desirable.

With the Bobrovnikoff method, the observer selects several nearby
comparison stars, some brighter and some fainter than the comet. Using a
magnification of 1.5 to 2 power per centimeter of aperture (to minimize the
apparent size of the comet):

(1) The telescope is defocused until the comet and stars have a similar
     apparent size.

(2) Go back and forth between a brighter and fainter pair of stars and
     interpolate the magnitude of the comet. (The interpolation method
     and example follow the description of the Morris method.)

(3) Repeat step (2) with several more star pairs.

(4) Take the average of the measurements in steps (2) and (3) and
      record it to the nearest 0.1 magnitude.

The Sidgwick or In-Out method is often used when a comet is too faint
to withstand any defocusing:

(1) Memorize the "average" brightness of the in-focus coma. This
     requires practice (and, unfortunately, this "average" often varies
     among observers).

(2) Defocus a comparison star to the size of the in-focus coma.

(3) Compare the defocused star's surface brightness with the memorized
     coma average brightness.

(4) Repeat steps (2) and (3) until a matching star is found or a
      reasonable interpolation can be made to the coma magnitude.

The Beyer method has fallen into disfavor because of the difficulty of
using it and because of its sensitivity to background sky brightness.

The Morris method matches the diameter of the moderately defocused comet
with a defocused star. The procedure is:

(1) Defocus the comet's head to obtain an approximately uniform surface
     brightness across it.

(2) Memorize the image obtained in step (1).

(3) Match the comet image _size_ with out-of-focus comparison stars. The
     stars will be more defocused than the comet.

(4) By comparing the _surface brightness_ of the defocused stars and the
      memorized comet image, estimate the comet's magnitude.

(5) Repeat steps (1) through (4) until an accurate magnitude estimate
     to the nearest 0.1 magnitude is made.

When the comet's apparent magnitude is between that of two comparison
stars, use the following standard interpolation method: Estimate the comet's
difference from the brighter star in step sizes of tenths of the difference
between the comparison stars. Then multiply the number of tenths by the
magnitude difference of the stars and add this product to the magnitude of
the brighter star. Rounded off to the nearest tenth, this is the comet's
apparent magnitude.*

---

* Example: Suppose comparison stars A and B are magnitudes 7.5 and 8.2,
            respectively. Their magnitude difference is 8.2 - 7.5 = 0.7.
            If the comet is 0.6 from A to B, then the estimated magnitude
            is 0.6 x 0.7 + 7.5 = 0.42 + 7.5 = 7.92 ≈ 7.9.

A set of charts showing Comet Halley's path from November 1985 through
May 1985 will be found in Part II. These charts, reproduced by the kind permission of the American Association of Variable Star Observers (AAVSO) and
the British Astronomical Association (BAA), should be used for all magnitude
estimates of the comet when it is brighter than magnitude 9.5, based on the
stellar magnitudes given on the charts. Stars that are obviously red should
not be used for comparison, however. It is simplest to just use the chart
on which the comet appears as a source of comparison magnitudes. Magnitudes
are usually marked to the right of the stars on the AAVSO charts and apply
to the star nearest the magnitude on the BAA charts. (Magnitudes on the BAA
charts are taken from AAVSO charts.) Magnitudes are given without a decimal
point, and an underlined magnitude was determined photoelectrically. Thus,
a star with 87 next to it is magnitude 8.7, and a star with 33 next to it is
photoelectric V magnitude 3.3. Magnitude estimates based primarily on underlined magnitudes should be underlined on the report form. This is very
important in interpreting the data.

At least three comparison stars should be used. Each observer's report
should quote the comet's estimated magnitude (or its average value from several observations made with the same instrument and magnification). Do not
apply any corrections to the value: follow the directions as given above and
avoid making mental compensations based on your perception of what "should"
be done for an accurate estimate. Write down the value immediately after the
estimate is made, along with the Universal Time to the nearest five minutes.
Do not rely on memory.

A dark sky background will decrease the likelihood of underestimating
the comet's brightness. Background sky brightness due to sources like city
lights, natural airglow, twilight, the Milky Way, and the zodiacal light*
varies with position across the sky whether or not a dark observing site is
used. Use extra care when making magnitude estimates when any of these background sources may affect your estimates. Make a note on the report form
whenever an estimate may have been affected by background sky light.

Telescope aperture and eyepieces used also affect magnitude estimates.
It is strongly recommended that observers planning to use two or more
instruments for magnitude determinations make estimates using both their
instruments when the comet is within reach of both. This recommendation is
made even for changing only eyepieces.

In fact, for a comet like Halley which is expected to get relatively
bright, the use of binoculars for magnitude estimates throughout much of the
apparition is quite appropriate. The empirical correction of observations
to a standard aperture of 67.8 mm will be required for all observations and
will be applied by IHW Archives users, not observers. (The correction is
more accurate for apertures close to the standard value, and reflecting tele-

---

* The times of the end of evening twilight and the beginning of morning twilight can be found in the U.S. Naval Observatory Astronomical Almanac, RASC
  Observer's Handbook, BAA Handbook, the Graphic Timetable of the Heavens
  and the Sky-Gazers Almanac for the current year. Tables 3-1 and 3-2 show
  zodiacal light visibility.

scopes need much less correction for aperture than refractors). Binoculars
with 50 mm and 80 mm apertures are commonly available. Also, to provide a
better tie-in with earlier apparitions, observers are encouraged to use
small aperture, long focus refractors for their observations.

Careful observers can make useful observations of "nuclear" magnitudes
(Whipple, private communication). Using a 15 cm or greater diameter telescope at high magnification, estimates of the brightness of the nucleus
should be made by comparison with stars of known magnitude plotted on the
charts in Part II. Estimates should be made on in-focus images of the
nucleus and stars. Several comparison stars should be used to make an
estimate accurate to ±0.1 magnitude. Nuclear magnitude estimates should
only be made when the seeing is very steady and the nucleus maintains a
stellar appearance at high magnifications. The nuclear magnitude should
be identified as such and reported in the "Notes" column on the Visual
Observation Report Form. Such estimates are useful in determining the
rotation rate and geography of "hot spots" on the nucleus.

As mentioned before, magnitude estimates are affected by aperture,
magnification, field of view, and background sky brightness. Because of
these influences it is imperative that coma diameter measurements be made
when magnitude estimates are made using the same instrument, also <u>without</u>
<u>using filters.</u> Small apertures are preferred for both magnitude and coma
diameter measurements.

One of four techniques should be used for coma diameter measurements.
Estimates based on the size of the field-of-view should <u>not</u> be used. If
the coma is elliptical, the length of both the long and <u>short</u> axes should
be measured.

The simplest but least accurate method for measuring the diameter of
the coma requires only an estimate of the coma size as a fraction of the
separation of two stars. The angular separation "S" of the two stars is
easily computed using their right ascensions (α_1 and α_2) and declinations (δ_1 and δ_2) in the formula

$$S = \cos^{-1} [\sin \delta_1 \sin \delta_2 + \cos \delta_1 \cos \delta_2 \cos(\alpha_1 - \alpha_2)]. \qquad (1)$$

The estimate should be made several times and the results averaged.

Another low-accuracy technique is to draw the coma on a detailed atlas
and measure its size using the scale of the atlas as a standard.

The coma diameter can be more accurately determined by using an
illuminated cross hair eyepiece (with a minimum of illumination) or an
eyepiece with an occulting blade covering about half of the field of view.
First, orient one cross hair east-west so that a star drifts precisely
along it when the telescope drive is off. Then simply time the interval
necessary for the coma to pass by the north-south cross hair. The same
technique works with an occulting blade eyepiece once the blade edge is
oriented north-south. It may be a good idea to begin with the coma out of
the field of view when the clock drive is shut off. This will remove ob-

servational bias in the position of the coma's leading edge. The diameter, d, in minutes of arc can then be computed with the formula

$$d = (1/4) \ t \cos \delta \qquad\qquad (2)$$

where t = interval in seconds;

δ = the comet's declination at the time of observation.

It is recommended that several interval measurements be made and the average value of these used.

The fourth and most accurate method requires a measuring reticle in the eyepiece or a filar micrometer. Once the scale is known*, the angular diameter may be computed immediately from the measurement. Eyepieces with built-in measuring reticles are available from a number of manufacturers. Micrometer construction is discussed by Worley (1961) and by Polman (1977). Some are commerically available.

Degree of condensation (DC) provides a description of the coma's intensity profile (i.e., the change in brightness with distance along a line through the coma centered on the central condensation). It ranges from 0 (diffuse image, no condensation, flat, smooth profile) to 9 (star-like image with stellar [point-like] intensity profile). Occasionally, comets develop a coma with a sharp edge like a planetary disk. Experienced observers will generally rate this DC = 9 since the coma is not diffuse at all. It should also be noted that a condensed comet need not have a central condensation. (A central condensation is a distinct disk in the coma.) Morris (1981a) discusses DC in detail. Whole number values of DC should be given; fractional values are not meaningful. Table 5-1 gives verbal descriptions, and Figure 5-1, supplied by J. Bortle, illustrates some values of DC.

Visual tail observations are of secondary importance to the IHW since photographic studies will generally obtain the tail length and position angle data required. Still, as a tie-in to data from past apparitions, these observations are useful. Observers should be aware that separating the gas and dust tails will be difficult because of the comet's low orbital inclination with respect to the ecliptic (18°).

For tails less than 10° in length, the tail's length should be estimated with respect to pairs of stars (as discussed for coma diameter observations). Again, estimates in terms of eyepiece or binocular fields-of-view are not recommended. For longer tails, Eq. (1) should be used with α_1, δ_1 referring to the comet's head and α_2, δ_2 referring to the end of the comet's tail. Background sky brightness can affect tail length estimates, just like coma magnitude estimates mentioned earlier. Take extra care when making these

---

* The scale can be determined by timing the passage of a star (with the telescope clock drive off) and using Eq. (2) and the number of divisions on the reticle passed by the star to determine the number of arc-minutes per division. This value varies with the focal length of the telescope used and will also change if the telescope focal length changes with, for example, temperature.

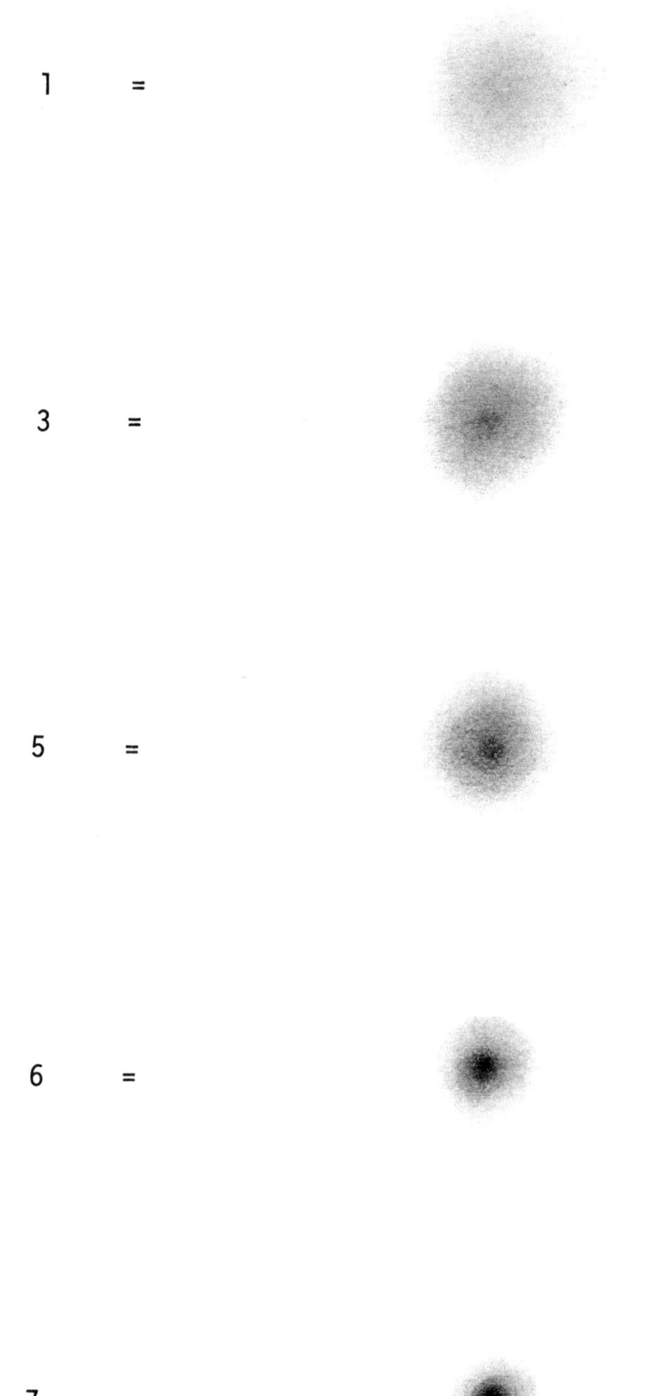

Figure 5-1. Drawings Showing Various Degrees of Condensation Supplied by John Bortle, W. R. Brooks Observatory

observations and make a comment in the notes section if there is a suspicion
that the estimate was affected by background sky light or if the tail is
curved.

TABLE 5-1

Degree of Condensation

DC	Description
0	Diffuse coma with uniform brightness, no condensation toward the center.
3	Diffuse coma with brightness increasing gradually towards the center.
6	Coma shows definite intensity peak at center.
9	Coma appears stellar.

Position angle (PA) is best determined by accurately plotting the position of the head and tail on a detailed star atlas and measuring the PA with
a protractor. This method can be accurate to ±5°. PA can also be estimated
by using the drift method to define east-west in an eyepiece and then
estimating, to the nearest half hour, what time on an imaginary clock face
the tail is pointing to. This method is only accurate to ±15°, which is
half the angle between any two hour-marks on a clock, and is not recommended
for this reason. A pointer attached to the outside of a cross hair eyepiece
can be used with a protractor or graduated piece of cardboard, wood, or
metal fixed to the telescope and zeroed on north (with respect to the cross
hairs and pointer; remember that with respect to gravity the position of
north in the field of view changes when any diagonal mirror or prism in a
telescope or eyepiece holder is rotated.) The cross hair is rotated to
the PA of the tail, and the value is read off the graduated circle. Due
north is defined as 0° PA, and PA increases through east, i.e., the trailing
side of an object allowed to drift out of the field of view.

When a star is seen through the tail, the PA of the star (subscript 2)
and thus the tail relative to the comet's nucleus (subscript 1) can be calcu-
culated from the known positions of the star and the nucleus with the formula

$$PA = \tan^{-1} \frac{\sin (\alpha_2 - \alpha_1)}{\tan \delta_2 \cos \delta_1 - \sin \delta_1 \cos (\alpha_2 - \alpha_1)} \qquad (3)$$

To determine the correct algebraic sign of the PA, determine the sign of
$[\sin (\alpha_2 - \alpha_1)]$. This will be the same as the sign of $[\sin PA]$. That is,

$$\text{sign} [\sin PA] = \text{sign} [\sin (\alpha_2 - \alpha_1)] \qquad (4)$$

Short gas and dust tails are generally straight. The PA of long,
curved dust tails should be measured at the root where it first leaves the
coma and at various positions in the rest of the tail. A measurement of
the distance from the nucleus must be included. This provides data on the
curvature of the tail.

Notes on the presence or absence of structures in the dust tail will
be useful as will reports on any changes seen in the tail and on the sizes
of the gas and dust tail roots relative to the coma. Mention of a bright
spine or dark "shadow of the nucleus" or hollow dust tail appearance can be
included in observational notes.

Observers are reminded that tail observations are very sensitive to sky
brightness and that moonlight or city lights can render such observations
useless. Twilight, the Milky Way, and the zodiacal light can also seriously
affect tail observations. Faint tails are sometimes more easily seen in large
aperture instruments.

The principal data acquired with inner coma observations are best rep-
resented in a drawing. Observations in twilight or moonlight can make fine
detail in the coma visible and simply staring for a while at the comet will
allow more structure to be seen. Halos, fans, rays, envelopes, jets, spines,
"nuclear shadows", and streamers should all be carefully drawn with their
correct sizes, shapes, orientations, and positions with respect to the
nuclear condensation. Don't rush. Soft lead pencils or charcoal drawing
supplies and paper stumps and erasers for smudging and erasing (available
from stationery and art supply stores) should be used to make negative (dark
comet on light sky) drawings of the comet. High magnification is generally
the best to use (long focal length refractors are superb for these observations), but several magnifications (or even several telescopes) can make for
a very accurate representation of the coma and/or whole comet. Filters may
make some details more visible, and their use should be recorded on the
report form. The _time_ of the drawing is an absolute necessity and the _scale
and orientation_ with respect to north and east must be shown. Measurements
of the vertex distance (distance from the central brightness maximum to the
vertex of the envelope as measured along the axis of the comet head) and of
the semi latus rectum values (the distances from the central brightness maximum to the envelope boundaries on both sides in the direction perpendicular
to the axis of the head) are very important to analysis of the nucleus
(see Fig. 5-2). The vertex distances and the semi latus rectum values for
internal envelopes should also be included. A written description of the
features on the drawing should be attached. It is best to start by sketch-
ing the positions of field stars from a star atlas and then going to the
telescope to fill in the cometary details. Observers can obtain valuable
practice by drawing various nebular objects visible in the sky (R. Shaffer,
private communication) as described by Eicher (1983).

High-quality drawings have value even with the advent of photography.
The eye is very good at resolving fine detail during moments of good seeing
and responds to an exceptionally wide range of intensity at one glance.
True representation of the relative intensity and position of the nuclear
condensation in the coma can yield valuable data on the state of excitation

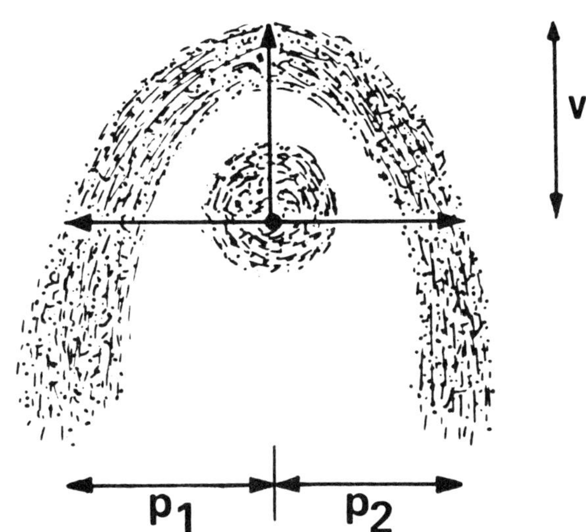

Figure 5-2. Depiction of Cometary Coma With Envelope. The Two Semi Latus Rectums May Not be Equal. In Such a Situation, They Both Should be Identified and Recorded Separately, Not Totaled Together. Courtesy, Dr. Fred Whipple

of the nucleus or of "hot spots" on its surface. Determination of rotation
and precession rates are possible using these data.

Whipple (in Brandt et al, 1981, in Marcus, 1981, and private communication)
urges amateurs to contribute drawings and measurements of halos, envelopes,
jets, and streamers observed in the coma. Good observations are extremely
valuable for determination of various characteristics of the nucleus. This is
an area where real contributions to understanding comet nuclei are possible and
where few observers are aiding the effort.

Visual comet observations yield valuable data and can be done by virtually any experienced observer. It cannot be emphasized enough, though, that
this area requires observational experience before reliable data are produced. Observers are urged to practice on all available comets and submit
their observations to the International Comet Quarterly or Association of
Lunar and Planetary Observers for evaluation and possible publication. A
formal test of IHW procedures is planned for a trial-run in 1984. Observers
are especially encouraged to participate in this activity.

Explanation of Visual Report Forms

<u>Chart No.</u> - The number of the IHW amateur manual's chart (<u>not page number</u>) used for comparison stars.

<u>Coma Dia.</u> - The coma diameter observed in minutes of arc. Give the long and short dimensions of an elliptical coma.

<u>Coma (Total) Magnitude</u> - The coma's estimated magnitude should be reported to the nearest 0.1 magnitude. If the magnitude is based on stars whose magnitudes are underlined, underline your estimate.

<u>Dark Adapted</u> - Indicate Y (yes) or N (no) if you were dark adapted when the comet observation was made.

<u>D.C.</u> - Give the degree of condensation of the coma.

<u>Faintest Star</u> - Give the magnitude of the faintest star visible to the naked eye (to within 0.5 magnitude) on the star chart in Part II containing the comet's position for the night of observation. M, T, C, or Z should be included with the stellar magnitude when moonlight, twilight, city lights, or zodiacal light (Table 3-1), respectively, interfere with the observation.

<u>Filter(s) Used</u> - List the filters used when the drawing was made.

<u>Instrument</u> - For <u>Aperture</u>, give the objective diameter in centimeters. <u>Type</u> describes the optical system (refractor, Newtonian, Cassegrain, Schmidt-Cassegrain, binoculars, etc.), and <u>f/</u> is the focal ratio of the instrument.

<u>Magnification</u> - Found by dividing the telescope focal length by the focal length of the eyepiece used for the observation.

<u>Magnification(s) Used</u> - List of magnification(s) used when the drawing was made.

<u>M.M.</u> - The magnitude estimation method used:

    B = Bobrovnikoff, S = Sidgwick, M = Morris

<u>Observer</u> - Each individual observer should use his/her own observing report form, complete with his/her name.

<u>PA</u> - Position angle of tail(s). For a curved tail give the distance from the nucleus where the measurement applies. Give the method used for determining it (plot, clock face, calibrated eyepiece). PA is defined to be 0° for due north and increases through 90° due east, 180° due south, and 270° due west. In the field of view, with the clock drive off, the last portion of an object drifting out of the field of view is the eastern piece. Thus, PA is well-defined even in circumpolar regions of the sky.

<u>Seeing</u> - Estimation of the seeing quality in seconds of arc or describe the seeing in some other standard manner.

<u>Site</u> - Give the name of your observing site used for the reported observa-
tion. If the site is not one listed on your Observer Index form, include
the longitude, latitude, and altitude. Longitude, latitude, and altitude
are available on topographic maps available at appropriate government
offices and some sporting goods and map stores. If these coordinates are
not available give the nearest town, village, or major landmark and its
distance and direction <u>from</u> the site.

<u>Tail Length</u> - Reported in degrees and tenths of degrees. Use two lines if
two tails are visible.

<u>UT Date and Time</u> - Local dates and times should be converted to UT as
explained in the Universal Time subsection, p. 4-6. Times given in UT
should be accurate to ± 5 minutes. A decimal date (e.g., Nov. 12, 12:00
UT = Nov. 12.50 UT) can be included if desired. Decimal dates should be
accurate to ±0.005 day.

<u>UT Start, UT End</u> - The beginning and ending times of the period when the
drawing was made.

VISUAL OBSERVATION REPORT FORM

Observer _______________________

UT Date and Time	M. M.	Coma (Total) Magnitude	Chart No.	Instrument Aperture	Type	f/	Magnifi-cation	Coma Dia.	D.C.	Tail Length	PA	Faintest Star	Dark Adapted	Site	Notes

DRAWING INFORMATION REPORT FORM

UT Date _________________________ Observer _____________________

Faintest Star ___________________ Site _________________________

Instrument Aperture _____________ Type ______________ f/________

Seeing ____

UT Start ________________________ UT End _______________________

Magnification(s) Used ____

Filter(s) Used ____

Features   Type   ID#   PA

            ______________   ______________   ______________

            ______________   ______________   ______________

            ______________   ______________   ______________

            ______________   ______________   ______________

            ______________   ______________   ______________

            ______________   ______________   ______________

            ______________   ______________   ______________

Indicate the orientation (north and east) in the drawing and the scale
(minutes of arc per millimeter).

Notes:

6. PHOTOGRAPHY

A heavy emphasis on photography is planned by professional astronomers
for Halley's Comet. Two separate discipline specialists will coordinate
photography of large and small scale phenomena (tails and near-nucleus,
respectively).

This does not mean, though, that amateurs cannot contribute. Weather,
telescope scheduling conflicts, observatory geographical coverage, and the
limited number of professional astronomers are reasons that amateurs can make
significant contributions with photographic observations. Observations of
the smallest scale phenomena require the long focal lengths usually available
only on large professional telescopes. Long focal length Cassegrain telescopes available to amateurs can provide valuable data at this scale. Other
professional planning is centered on telescopes with fields-of-view of about
$5° \times 5°$, and many professional instruments are limited to comet elongations
of $30°$ or more from the sun. Thus, there are also niches for amateur astronomers to fill by providing supplementary large-scale observations.

In order to provide useful data, it is advisable to standardize the
photographic emulsions, processes, and techniques as much as possible. Such
standardization will make data analysis simpler.

Black and white (B/W) photography will be emphasized for the Halley
Watch. Color photography can provide spectacular and useful pictures, but
the variations in emulsions and processes work together to create ambiguity
during the intercomparison of the data.

Ideally, a photographic emulsion should be very fine-grained to retain
resolution to the seeing limit and very sensitive to decrease the data
acquisition time. This combination has been the goal of film manufacturers
for years. At this time, only special preprocessing (called hypersensitizing) of some emulsions allows adequate speed to be achieved. Otherwise,
coarser-grained emulsions must be used for highest sensitivity, and fine-
grained, less sensitive emulsions are used in other situations.

IHW photographers are encouraged to use hypersensitized, fine grain
emulsions like Eastman Kodak's 2415 Technical Pan film or its equivalent
whenever possible. It has been found that bathing this emulsion (and others)
in warm forming gas (a nonexplosive mixture of hydrogen and nitrogen) for
several hours before exposure significantly increases the sensitivity of
the emulsion. Amateurs wishing to hypersensitize their film using this
or other processes should study the references given at the end of this
section for details of the processes.

Amateurs lacking hypersensitizing equipment should use high-speed (ISO
400 or so) moderately grainy films available from Kodak, Ilford, Agfa-Gaevert,
or others. In some situations moderate-speed films (ISO 100-125) with finer
grain may prove useful, including nonhypered 2415, Kodak Plus X Pan, Ilford
FP 4, and similar products.

Postexposure processing should be done as recommended by the film manufacturer. Low or normal contrast development should be used to bring out coma and tail details. The use of high-contrast developers like MWP-2, Kodak D-19, and similar types provide the necessary speed and contrast to show the overall extent of the comet.

Because the comet is a moving target and tail structure is relatively faint, long exposures will be necessary to record detail. <u>Correcting for the comet's apparent motion across the sky</u> is absolutely vital. Even relatively short exposures will require accurate guiding on the comet since it is moving with respect to the stars.

There are several methods of correcting for a comet's motion across the sky, listed in order of decreasing accuracy:

1) The most accurate method is to compute the differential motions in right ascension and declination and then drive the telescope at those rates with respect to the clock drive. A microprocessor control makes this easy, but rate-calibrated drive correction motors on the telescope axes will also do the job. Close to the horizon, differential refraction by the atmosphere can move the comet around the field of view in an irregular manner. Observers using this method should regularly check the tracking.

2) Compute the angular motion and align a filar micrometer cross hair along the position angle of the motion. At predetermined intervals, offset the cross hair in the direction opposite the comet's motion and recenter the guide star. The correction rate depends on the angular rate of motion. This method is discussed in a pair of papers by Lines (1973a, b).

3) Guide on any nucleus or distinct central condensation. Cross hairs or a tapered pointer are good for maintaining accurate centering (Fig. 6-1).

4) Compute the angular motion and align cross hairs tangent to the coma so that motion is directed along the diagonal of the opposite quadrant (Fig. 6-1).

5) The least accurate method is to center cross hairs on the coma and attempt to keep the cross hairs on the same point in the coma (Fig. 6-1).

It should be obvious that off-axis guiding will not be effective since the photographic target must be used to guide on in this case. Guiding methods (2), (3), (4), and (5) all require a co-aligned guide telescope.

Tail photography is one area where amateurs can make useful contributions. Halley's visual tail is not expected to exceed about 30° in length because of the poor circumstances of this apparition. This will conveniently fit the length of a standard 35 mm camera frame with a 50 mm focal length lens. Such lenses are typically sharper at the corners of the field when stopped down one to two f/stops. Each observer is requested to <u>obtain nega-</u>

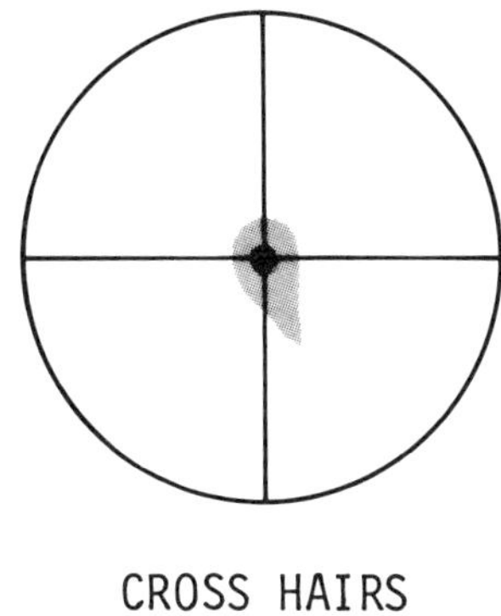

CROSS HAIRS

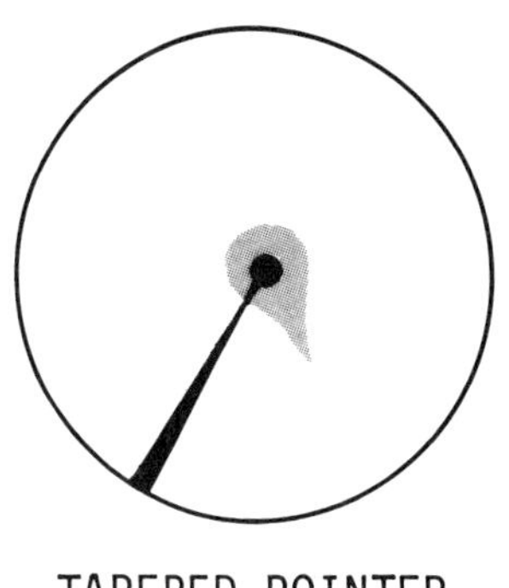

TAPERED POINTER

3. ON CONDENSATION

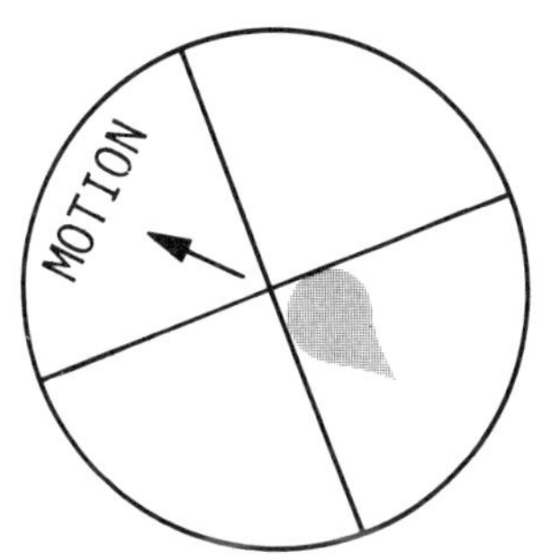

4. TANGENT CROSS HAIRS

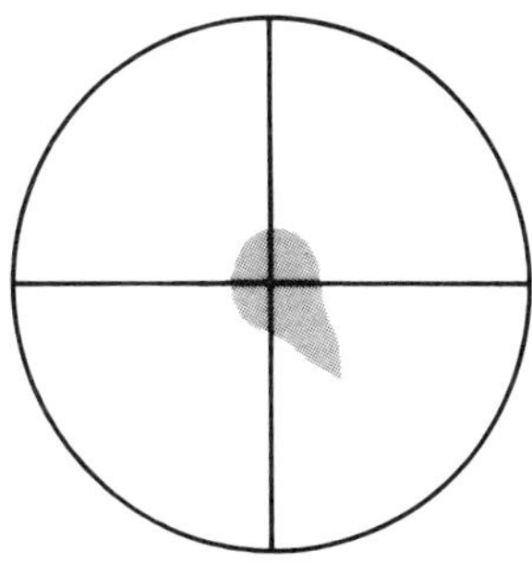

5. CROSS HAIRS ON COMA

Figure 6-1. Methods 3, 4, and 5 for Accurately Guiding a Camera on a Comet

tives of a two-minute guided exposure centered on the belt of Orion and a 20-minute exposure centered on either M31 (northern hemisphere observers) or M83 (southern hemisphere observers) for each wide field lens used. These will be used for photometric calibration and scaling purposes (since manufacturers' claimed focal lengths vary as much as ±10% off the production line). At least one calibration exposure of M31 or M83 should be made on each roll of film used. A single negative of Orion obtained early in the apparition with each lens planned for comet use will be sufficient.

Filter photography of the tail will provide valuable data as well as useful historical comparisons. Photographic emulsions in 1910 were blue-sensitive and not panchromatic, so they preferentially recorded the ion tail.

The ion tail can be isolated optically using an interference filter transmitting CO^+ wavelengths from 4100 Å to 4600 Å and excluding neutral molecular lines and scattered solar continuum. The dust tail is separable using a filter transmitting a "clean" continuum wavelength.

For tail photography, a sequence including unfiltered, blue, and orange images is suggested. Suitable combinations of glass or gelatin filters placed in front of the camera lens or immediately before the film plane will allow tolerable isolation of the tail components. The blue filter's primary transmission band should be centered at 4400 Å with a width at the 50% transmission points of 900 Å and peak transmittance of at least 63%. The orange filter should have a sharp cut-on beginning transmission at 5400 Å and exceeding 90% transmission at wavelengths of 6500 Å and greater. Kodak gelatin filters 47A and 21, respectively, satisfy these specifications. The combination of 2B with 47A is even better in the blue. Glass filters matching these specifications are available from a variety of manufacturers - check with a camera store.

Blue images emphasize the ion tail of the comet while orange images emphasize the dust tail and its structure. A sequence (possibly a movie) made from these images through the apparition may be very instructive. Several exposure sets each night coupled with those taken at other longitudes will allow temporal changes to be detailed.

When the observing window for the comet is short (for example, when comet rises shortly before twilight starts), it may not be possible to obtain two filter photographs of the tail. In such a situation, a color photograph can be used to record both tails simultaneously with the advantage that observing conditions, guiding, and atmospheric refraction effects are identical. In the darkroom, photographic subtraction will allow isolation of the tails. Original color transparencies used for subtraction purposes should not be push processed.

For the copying steps that are necessary in the method of photographic subtraction, the copies should have low contrast, even development, and equal image scales (avoid refocusing at different stages) for proper subtractions. Make an enlarged or contact B/W negative of the original color positive transparency, an enlarged or contact B/W negative of the transparency projected through a blue Wratten 47B filter, and an enlarged or contact B/W negative of the transparency through a red Wratten 25 filter. The two "filtered" negatives should have comparable densities. From the filtered negatives, make B/W contact positives such that when a positive is placed

emulsion-to-emulsion with its original negative, an even grey appearance
over the comet and sky is produced (stars may not exactly cancel because
their images may be saturated on the copies).

To isolate the ion tail, make a print with the unfiltered negative and
the "red positive" facing each other emulsion-to-emulsion with stars aligned
to overlap. To isolate the dust tail, make a print with the unfiltered
negative and the "blue positive" emulsion-to-emulsion.

Moderate scale photographs (with fields-of-view approximating professional instruments) may provide useful supplementary data to that obtained
by the Large Scale Phenomena net. This could be especially true if clouds
interfere with professional observations or better time resolution is needed.
The calibration photograph discussed below with narrow-angle photography
should also be made with instruments operating at moderate scales.

Narrow-angle coma photography can be accomplished with telescope apertures of 15 cm or more and focal lengths of 2500 mm or more. A series of
bracketed exposures that vary by a constant factor (say, 2, so the exposure
doubles from one to the next) will provide good results in spite of unknown
amounts of atmospheric attenuation and if calibration photos (discussed below) cannot be obtained. A series will also better portray the full dynamic
range of intensity from the inner coma to the outer coma as well as details
such as jets, spiral structure, or small scale splitting of fragments from
the nucleus. Photos obtained through polarizing filters (military surplus,
not camera store types due to their inefficiency) at several known position
angles may provide interesting data, but such experiments should not be
emphasized over regular high-resolution photography. Narrow-angle photography is an area where amateurs' contributions could lead to a much better
understanding of nuclear phenomena because of the constant monitoring
possible with relatively large numbers of dedicated observers.

In order to make meaningful use of narrow-angle data, a ten-minute
exposure of M31 (northern hemisphere observers) or M83 (southern hemisphere
observers) should be obtained for each narrow-angle photographic configura-
tion used. At least one such photograph should be made on each roll of
film used to establish photometric calibration for all the negatives. This
will also provide the necessary scaling data. Remember that optical systems
which vary the spacing of the primary and secondary elements also vary the
focal length and image scale in the process. This is a characteristic of
most commercial Schmidt-Cassegrain and Maksutov telescopes. It may prove
necessary to obtain scaling photos at different temperatures if the focus
(i.e., primary-secondary separation) shifts with temperature.

The calibration photographs requested for all three types of photography
are important to the data analysis. While this manual mentions only the
use of roll film, the IHW recognizes that sheet film and glass plates are
also commonly used. Since both pre- and post-exposure photographic processing
affect the photometric interpretation of the data, it is important that the
calibration images be processed in the same manner as the comet images.
This is easily done with roll films and is necessary, if less convenient,
with sheet film and plates.

There are several useful contributions that amateurs can make using
photography. Careful effort is necessary for them to be of value.

Astrophotographic Emulsion Treatment Bibliography

Brown, G. P., G. T. Keene, & A. G. Millikan, "An Evaluation of Films for Astrophotography," Sky and Telescope, May 1980, p. 433.

Ernest, O., "Deep Sky Photography with Cooled Emulsions," Sky and Telescope, Mar. 1973, p. 189.

Everhart, E., "Adventures in Fine-Grain Astrophotography," Sky and Telescope, Feb. 1981, p. 100.

Everhart, E., "Hypersensitization and Astronomical Use of Kodak Technical Pan Film 2415," AAS Photobulletin, 1980, No. 2 (Consec. No. 24), p. 3.

Healy, D., "Experiments with Gas-Hypered Film," Sky and Telescope, Feb. 1981, p. 174.

Hollars, D. R., "Temperature Effects on Photographic Sensitivity," AAS Photo-bulletin, 1971, No. 2 (Consec. No. 4), p. 18.

Iburg, B., "The Shoot Out: Cold Camera vs. Gas Hypering," Astronomy, Nov. 1981, p. 61.

Jones, E. L., "A Cold Camera That Needs No Vacuum," Sky and Telescope, Aug. 1975, p. 122.

Lapointe, M., "Which Films Are Worth Hydrogenating?" Sky and Telescope, May 1978, p. 401.

Marling, J. B., "Gas Hypersensitization of Kodak Technical Pan Film 2415," AAS Photobulletin, 1980, No. 2 (Consec. No. 24), p. 9.

Santos, J. M., "Hypersensitizing Films for Astrophotography," Sky and Telescope, Nov. 1970, p. 322.

Schoening, W. E., "Nitrogen Plate Baking at Kitt Peak," AAS Photobulletin, 1975, No. 3 (Consec. No. 10), p. 18.

Sky and Telescope, "Notes on Gas Hypersensitizing," Feb. 1981, p. 176.

Smith, A. G., "New Trends in Celestial Photography," Sky and Telescope, Jan. 1977, p. 24.

Smith, A. G. & H. W. Schrader, "Balanced Hypersensitization of a Fast Reversal Color Film," AAS Photobulletin, 1979 No. 2 (Consec. No. 21), p. 9.

Stong, C. L., "The Amateur Scientist," Scientific American, Aug. 1969, p. 125. "Color photographs of the night sky are made by refrigerating the film."

Stong, C. L., "The Amateur Scientist," Scientific American, Dec. 1973, p. 122. "A cold camera for astronomical photography...".

Wallis, B. D. & R. W. Provin, "On the Road to Better Astronomical Photographs," <u>Sky and Telescope</u>: Apr. 1977, p. 314; May 1977, p. 399; June 1977, p. 484.

Walker, P. E., "Hypersensitization of Kodak Technical Pan Film 2415 by Bathing in Silver Nitrate Solution," <u>AAS Photobulletin</u>, 1980 No. 2 (Consec. No. 24), p. 7.

Wiseman, J. D., Jr., "Cooled-Emulsion Photography Without a Vacuum," <u>Sky and Telescope</u>, Feb. 1969, p. 118.

Explanation of Photographic Information Report Form

<u>Duration</u> - Give the exposure time in minutes and tenths of minutes for long exposures and in seconds for short exposures.

<u>EFL</u> - Give the effective focal length of the photographic method used.

<u>Faintest Star</u> - Give the magnitude of the faintest star visible to the naked eye (to within 0.5 magnitude) on the star chart in Part II containing the comet's position for the night of observation. M, T, C, or Z should be included with the stellar magnitude when moonlight, twilight, city lights, or zodiacal light (Table 3-1), respectively, interfere with the observation.

<u>Film Name</u> - The manufacturer and film type should be listed as well as the ISO (ASA/DIN). If the film has been hypersensitized, give the method (dry nitrogen, forming gas, silver nitrate rinse, alcohol rinse, etc.), the temperature of the solution, and the duration of the soak. For cooled emulsion photography, give the temperature at which the exposure was made. Indicate the applicable temperature scale. The film processing method should include developer, temperature, and time (or write "commercial" if processed professionally).

<u>Filter</u> - List the color and designation of any filter used. Examples are Yellow, Wratten 12; Light Blue, Schott BG 38; etc. For Schott, Corning, and similar filters, include the thickness of the glass.

<u>Guiding</u> - Indicate the guiding method used (see this section for detailed descriptions).

<u>Instrument Focal Length and f/</u> - Focal length in mm and focal ratio (f/#) of the instrument used.

<u>Negative Number</u> - Give the number of the negative to which the exposure details apply.

<u>Notes</u> - Include comments on special circumstances, unusual events, exceptions, and deviations recognized during the observation or in the methods used to make the observation.

<u>Observer</u> - Each individual observer should use his/her own observing report form, complete with his/her name.

<u>Photographic Method</u> - Indicate the type based on the following light paths:

    Principal Focus (PF): telescope objective - film
    Negative Projection (NP): telescope objective - negative lens - film
    Eyepiece Projection (EP): telescope objective - eyepiece - film
    Afocal (A): telescope objective - eyepiece - camera lens - film

$\underline{Site}$ - Give the name of your observing site used for the reported observation. If the site is not one listed on your Observer Index form, include
the longitude, latitude, and altitude. Longitude, latitude, and altitude
are available on topographic maps available at appropriate government
offices and some sporting goods and map stores. If these coordinates are
not available give the nearest town, village, or major landmark and its
distance and direction <u>from</u> the site.

<u>UT Date</u> - Give the date based on Universal Time.

<u>UT Date Range</u> - Give the first and last UT date included for the photographs
described on the report form.

<u>UT Start</u> - Give the Universal Time of the beginning of the observation.

PHOTOGRAPHIC INFORMATION REPORT FORM

UT Date Range _____________________ Observer _____________________

Instrument Focal Length _____________ f/ _____________

Photographic Method: PF __   NP __   EP __   A __   EFL = _______ mm

Film Name _______________________ ISO (ASA/DIN) _______________

Hypersensitized in _______________________ at ____ °C/°F for ____ hours

Emulsion cooled to ____ °C/°F

Developed in _______________________ at ____ °C/°F for ____ minutes

Guiding: Computed ___   Micrometer ___   On Condensation ___

        Tangent X-hairs ___   X-hairs on Coma ___

Exposures

Negative Number	UT Date	UT Start	Duration	Filter	Faintest Star	Site

Notes:

Submit contact prints or duplicate slides with your name on them to the
Photography Recorder.

The precise measurement of the position of Halley's Comet with respect to background stars is very important to orbit computations, ephemeris predictions, and nucleus modelling efforts. An accurate ephemeris (list of predicted positions at particular times) is an absolute necessity for the navigation of the spacecraft that will make close flybys of the comet and to radio astronomers who must point at and track the comet for long periods without visually guiding on it. Because few professional astronomers are doing astrometry, the Astrometry Discipline Specialist Team is seeking additional astrometric observations from amateurs.

A measurement accuracy of less than one second of arc for the comet is necessary. Measurements should be made with a measuring engine capable of measuring relative positions to an accuracy of about 1 micrometer. These parameters suggest that a camera focal length of greater than 200 mm will, in principle, give sufficient image scale for accurate measurements. However, a longer focal length in the range of 1 m to 2 m is strongly recommended. Penhallow (1978) discusses the design, construction, and use of an astrometric reflector. However, a telescope need not be a dedicated astrometric instrument to provide valuable astrometric data.

Astrometric photography should use the same techniques as cometary photography. The standard methods of emulsion development after exposure and, if desired, hypersensitization before exposure* can be used.

Guiding methods (1) or (2) described in the Photography section are the best to use when making astrometric photographs. Exposures should be just long enough to show the central condensation of the coma and some reference stars. Such an unspectacular comet photograph is the most easily measured since the stars are less trailed and the center of the condensation is more easily seen. <u>The time of the middle of the exposure must be known to the nearest second.</u>

Measurements should be made with an accurate measuring engine. These can be home-built (Penhallow, 1978, and Everhart, 1982) if desired. Local observatories or universities may have a measuring engine that could be made available for Halley measurements. No matter what measuring engine is used, it must be free of play and periodic errors in its screws.

Experienced users of measuring engines have developed "tricks" to improve the accuracy of their measurements. Original negatives, emulsion side up, are always measured. Contact prints or enlargements are never used. If the negatives are on film (and not glass plates), they should be sandwiched between two pieces of plate glass during measurement. At least three stars and the comet must be measured on the negative, and the stars should be evenly distributed around the comet. Faint stars with small images are easier to measure than bright stars with large images. The

---

* Hypersensitizing the emulsion may not be very helpful since the exposures
  for astrometry are often short enough that low intensity reciprocity
  failure (which hypersensitization helps correct) is not a problem.

negative should be measured in one orientation and then rotated 180° and
remeasured to cancel out several kinds of error. Finally, all images should
be measured with the measuring engine screw traveling in the same direction.

The mathematical reduction method that follows is designed for three
stars and is the one suggested by Marsden (1982) and Marsden and Roemer in
Wilkening (1982). Tatum (1982a) gives a very complete and readable discussion of comet astrometry. For improved accuracy, four to eight evenly
distributed reference stars should be measured with the comet and the
method of least squares used (see, for example, the chapter by F. Schmeidler
in Roth, 1975, p. 206) to obtain a solution for the comet's position.

This procedure should be followed for each reference star (see Fig. 7-1):

(1) Identify a reference star on the negative and find it in the
     Smithsonian Astrophysical Observatory Star Catalog (1966). Using
     the listed values of proper motion in right ascension and declination
     update the right ascension (α) and declination (δ) of the star from
     1950.0 to the epoch (date) that the negative was taken. Do not pre-
     cess the positions to the current epoch however. The resulting α
     and δ are referred to the 1950.0 equinox.

(2) Adopt a value for the position on the sky of the center of the
     negative in right ascension (A) and declination (D). Great
     accuracy in these values is not necessary. The same A and D for
     the center should be used for each reference star and the comet.

(3) Compute the following values:

$$H = \sin \delta \sin D + \cos \delta \cos D \cos (\alpha - A) \qquad (1)$$

As a check on H, note that its value should be approximately 1.

$$\xi = \frac{\cos \delta \sin (\alpha - A)}{H} \qquad (2)$$

$$\eta = \frac{\sin \delta \cos D - \cos \delta \sin D \cos (\alpha - A)}{H} \qquad (3)$$

(4) Set the focus of the measuring engine on the comet's image and
     don't change it thereafter. Measure the star coordinates x and y
     on the negative with the measuring engine. If the stars are
     distinctly streaked from compensating for the comet's motion the
     ends of the trails should be measured and the average value used
     for the stars' positions at mid-exposure. The zero point for
     measurements on the negative can be anywhere on the negative but
     must be the same point for all measurements of the reference stars
     and comet. The units of x and y can be millimeters, inches, or
     whatever is convenient. Remember that all star and comet measure-
     ments should be made with the measuring engine screw traveling in
     the same direction.

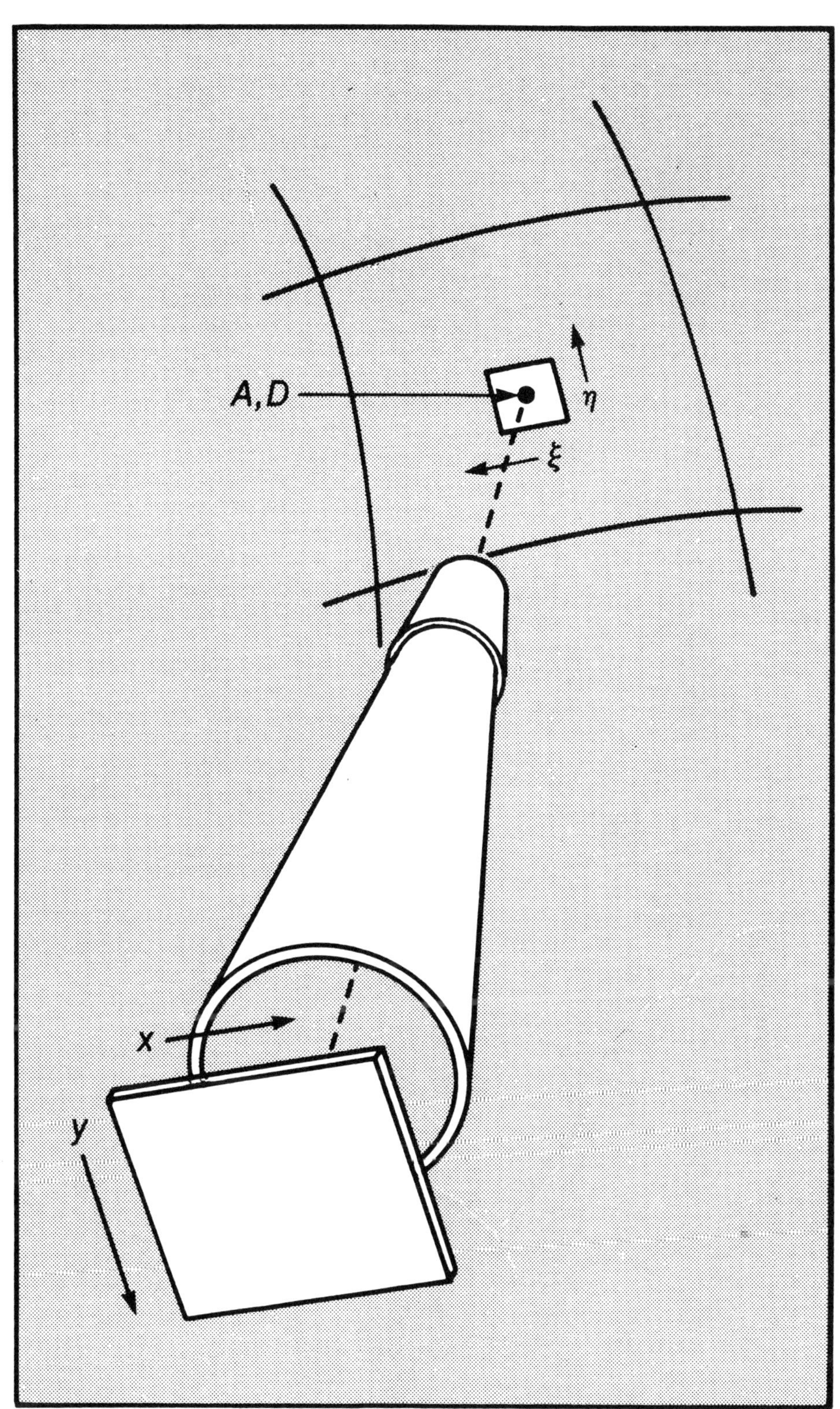

Figure 7-1. Quantities Used in Measuring a Comet's Position on a Photograph are Shown Schematically Here. The x, y Coordinates of the Comet and Reference Stars are Perhaps Expressed in Inches or Millimeters. They are Connected by Equations to Dimensionless "Standard Coordinates" ξ, η on the Sky. The x, y Origin Might be Near the Corner of the Plate (It Matters Not Where), But ξ, η are Each Zero at the Right Ascension and Declination of the Photograph's Adopted Center. Reproduced by Permission of <u>Sky and Telescope</u>

(5) Adopt a value for the focal length (F) of the telescope and
     express it in the same units used for the quantities x and y.
     The value of F does not need to be known with great accuracy, but
     the same value should be used for each reference star.

(6) Generate the following equations:

$$\xi - \frac{x}{F} = ax + by + c \tag{4}$$

$$\eta - \frac{y}{F} = a'x + b'y + c' \tag{5}$$

where a, b, c, a', b', c' are unknown plate constants.

Repeating steps (1) through (6) for three reference stars will generate
three pairs of equations (4) and (5), i.e., six equations for the six unknown plate constants. Using standard algebraic techniques commonly found
in algebra books, the values of the plate constants may be found. As a
rough check it should be found that a is approximately equal to b' (a $\approx$ b')
and that b is approximately equal to -a' (b $\approx$ -a').

Now measure the position x" and y" of the comet on the negative. Using
x" and y", the computed values of a, b, c, a', b', and c' and the adopted
value of F, solve the equations:

$$\xi'' = \frac{x''}{F} + ax + by + c \tag{6}$$

$$\eta'' = \frac{y''}{F} + a'x + b'y + c' \tag{7}$$

The comet's rectangular coordinates on the sky are ξ'' and η''.

Now solve the following equations in the order given to find the 1950.0
astrometric coordinates of the comet. The coordinates A and D adopted
earlier for the plate center are used again:

$$\Delta = \cos D - \eta'' \sin D \tag{8}$$

$$\Gamma = \sqrt{(\xi''^2 + \Delta^2)} \tag{9}$$

$$\alpha'' = A + \tan^{-1} \frac{\xi''}{\Delta} \tag{10}$$

$$\delta'' = \tan^{-1} \frac{\sin D + \eta'' \cos D}{\Gamma} \tag{11}$$

No attempt should be made to correct measurements for the effect of parallax.

Precise astrometric positions can only be achieved by experienced and disciplined observers. Amateur astronomers who wish to contribute to the International Halley Watch Astrometry Net should begin their observing programs as soon as possible to refine their techniques on other comets in advance of the 1985-86 apparition of Halley. There are only a very few amateur and professional astronomers who regularly contribute astrometric positions of comets to the Central Bureau for Astronomical Telegrams (see Appendix A for the address). Hence this is an under-represented observing discipline well-suited to serious amateurs who enjoy the challenge of doing precision work.

Accurate measurements will directly contribute to making precise ephemerides for Halley's Comet. This, in turn, will directly and indirectly help investigations into the nature of Comet Halley.

Explanation of Astrometric Data Report Form

Comet Image - Note whether or not the measured comet image was diffuse or
had an obvious central condensation. The discussion of degree of condensation in the Visual Observations section may be instructive.

Duration - Give the exposure time in minutes and tenths of minutes for long
exposures and in seconds for short exposures.

EFL - Give the effective focal length of the photographic method used.

Film Name - The manufacturer and film type should be listed as well as the
ISO (ASA/DIN). If the film has been hypersensitized, give the method (dry
nitrogen, forming gas, silver nitrate rinse, alcohol rinse, etc.), the
temperature of the solution, and the duration of the soak. For cooled
emulsion photography, give the temperature at which the exposure was made.
Indicate the applicable temperature scale. The film processing method
should include developer, temperature, and time (or write "commercial" if
processed professionally).

Guiding - Indicate the guiding method used (see the Photography section for
detailed descriptions).

Instrument Focal Length and f/ - Focal length in mm and focal ratio (f/#)
of the instrument used.

Negative Number - Give the number of the negative to which the exposure
details apply.

Notes - Include further explanation for the Comet Image classification, if
necessary, and comments on special circumstances, unusual events, exceptions,
and deviations recognized during the observation or in the methods used to
make the observation or to do the data reduction.

Observed α, Observed δ - Give the right ascension (α) to two decimal places
in seconds of time and declination (δ) to one decimal place in seconds of arc
computed for the comet's position for the UT date and time of mid-exposure.

Observer - Each individual observer should use his/her own observing report
form, complete with his/her name.

Photographic Method - Indicate the type based on the following light paths:

    Principal Focus (PF): telescope objective - film
    Negative Projection (NP): telescope objective - negative lens - film
    Eyepiece Projection (EP): telescope objective - eyepiece - film
    Afocal (A): telescope objective - eyepiece - camera lens - film

Site - Give the name of your observing site used for the reported observation. If the site is not one listed on your Observer Index form, include
the longitude, latitude, and altitude. Longitude, latitude, and altitude
are available on topographic maps available at appropriate government

offices and some sporting goods and map stores. If these coordinates are
not available, give the nearest town, village, or major landmark and its
distance and direction <u>from</u> the site.

<u>UT Date</u> - Give the date based on Universal Time.

<u>UT Date Range</u> - Give the first and last UT date included for the photographs
described on the report form.

<u>UT of Mid-Exposure</u> - Give the Universal Time of the <u>middle</u> of the photographic
exposure, accurate to 1 second.

ASTROMETRIC DATA REPORT FORM

UT Date Range ___________________________   Observer ___________________________

Instrument Focal Length ________________   f/ ________________

Photographic Method: PF ___   NP ___   EP ___   A ___   EFL = _____ mm

Film Name ________________________________   ISO (ASA/DIN) ________________

Hypersensitized in ___________________________ at ____ °C °F for ____ hours

Emulsion cooled to ____ °C °F

Developed in ___________________________ at ____ °C °F for ____ minutes

Guiding: Computed ___   Micrometer ___   On Condensation ___

          Tangent X-hairs ___   X-hairs on Coma ___

Exposures

Nega-tive #	UT Date	UT of Mid-Exposure	Dura-tion	Site	Computed α	Computed δ	Comet Image

Notes:

Submit this form to the Coordinator for Amateur Observations.

Spectroscopic studies of astronomical objects have led to much of our understanding of the physical nature of these objects. Few amateurs have attempted to duplicate professional work in this area. The Photography section of this manual contains useful information for observers planning spectroscopic observations. Black and white emulsions should be used for these studies.

There are three spectroscopic methods which amateurs can easily use that will yield data to supplement professional observations of Halley's Comet. "Objective" spectroscopic observations require a disperser (prism or diffraction grating) to break down incoming light into its component wavelengths and a camera lens to focus the spectrum on the film. "Nonobjective" spectroscopy requires a disperser, film holder, and telescope. A slitless spectrograph can be assembled which mimics observatory spectrographs in most of its features. No slits are necessary for any of these methods. The observer need only guide the camera or telescope on the comet for good results.

Prisms are easier to obtain than gratings and put all the light into one spectrum. Their disadvantages are that they absorb some light and do not disperse (spread) the light linearly; i.e., spectral lines are more crowded at the red end of the spectrum than at the blue end. Also, extra care is necessary to point them at the object of interest since they refract the rays as well as disperse them. A separate finder telescope or pointer is useful.

Diffraction gratings have a series of parallel grooves ruled on them, several hundred to the millimeter, which act to spread the light into a spectrum. They come in reflection and transmission varieties, and the best ones are generally very efficient. Dispersion depends on the number of grooves/mm and is nearly linear as a function of wavelength. Reflection gratings are hard to point at the object of interest, and a separate finder or pointer is almost a necessity. With transmission gratings, the object and its spectrum (if it's bright enough) can be seen in the camera's finder.*
The recorded spectrum is easier to analyze if a spectrum and its source are photographed together.

The disadvantages of gratings are that the grooves are delicate and easily damaged and that they produce a large (actually infinite) number of individual spectra, called orders, which make any particular order less intense than if all the light were going into a single spectrum. A solution to this problem is found by adjusting the groove shape. Most of the light can be directed into one order and is called "blazing" the grating. Orders are numbered increasing outward on both sides from the zero order (which is the direct, undispersed image of the object). Spectra generally get fainter with increasing order number except for the blazed order which is brightest

---

* Experience shows that if the spectrum of a first magnitude star is visible when the star is viewed through a transmission grating with the naked eye, the grating is efficient enough for use in spectroscopic observations. Unfortunately, the cheapest gratings made of plastic film do not meet this criterion.

by design. Orders starting with the second order and increasing have greater and greater overlaps on each other.

Objective prism spectroscopy by amateurs is discussed by Waber and McPherson (1967) and Patterson and Michaud (1980). Basically, a 25° to 60° prism is placed in front of a camera lens (50 mm to 200 mm focal length) so only light going through the prism reaches the lens. The prism and lens combination are chosen to ensure that the spectrum fits in the camera field-of-view. The prism should be oriented at the angle of minimum deviation, which means that the angle of incidence of light into the prism should equal the angle of emergence of light from the prism. The angles of incidence and emergence are measured with respect to the perpendiculars from the prism faces.* Spectra should be obtained with the dispersion perpendicular to the direction of the comet's tail. During the exposure, the spectrograph should be guided on the comet using one of the methods described in the Photography section.

Analysis of spectral data will be performed by users of the IHW Archives. Observers may want to determine wavelengths for their own information. To compute the wavelengths of spectral lines, detailed information on the glass in the prism is required. It is easier to determine wavelengths using a calibration curve based on known wavelengths and their position on the film (Schmiedeck, 1979).

The two simple grating spectrograph designs useful to the amateur (discussed by Edberg, 1982) both work best if the grating is blazed for visual wavelengths (4000 Å to 7000 Å) in the first order. With objective grating spectroscopy, a grating with 300 to 600 grooves/mm is placed in front of a camera lens whose focal length is 35 mm to 100 mm (for standard 135 film). Only light passing through the grating should reach the lens. The spectrum should be oriented so that it is dispersed perpendicularly to the comet tail, and the zero order and first order spectra should fit in the field of view. Some observers may wish to obtain the second order in addition. The efficiency of the grating, film speed, focal ratio (f/number) of the optical system, sky brightness, and comet brightness play a part in determining the proper exposure. A minimum of five minutes on fast black and white films or on hypersensitized fine grain films is recommended (see the section on Photography).

The curious observer can identify the wavelengths of spectral lines by using the formula

$$\lambda = \frac{n}{d} \; \frac{L}{\sqrt{L^2 + F^2}} \tag{1}$$

---

* To see this effect, look through a prism at an object and rotate the prism slowly on its axis. The object will appear to move, come to a halt, and move back in the opposite direction. The angle of minimum deviation is the orientation at which the image motion stops.

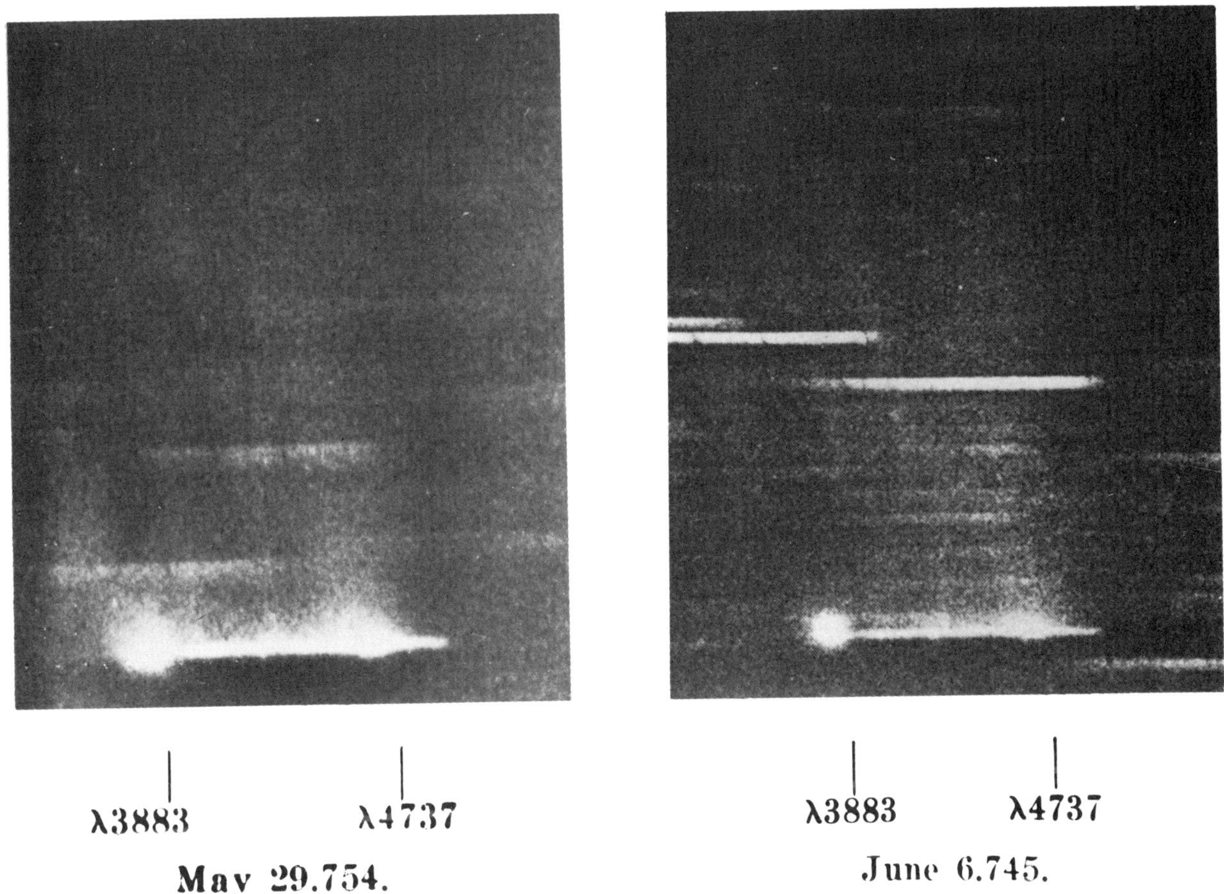

Figure 8-1. Objective Prism Spectra of Halley's Comet in 1910. These Examples are Reproduced from Bobrovnikoff's Classic Work (1931). Lick Observatory Photographs.

where $d = \dfrac{1}{\text{\# grooves/mm}}$ of the grating;

      n = order number of the spectrum used for measurement;

      L = the distance on the film or plate from the zero order image to the spectral line in millimeters;

      F = the focal length of the lens used in millimeters.

With nonobjective spectroscopy, the prism or grating is placed between any telescope objective (used as a light collector) and a film holder (e.g., camera body) to hold the film. A camera lens is unnecessary as the telescope acts to collect and focus light which is dispersed on the way to the film plane. It is important to focus on the spectrum and not on the source in this system since the disperser introduces coma, astigmatism, and field curvature. A slow optical system is preferred because the smaller cone angle of the converging beam decreases wavelength uncertainty in the spectrum.

This technique has been used with great success with telescopes as large as four meters for the detection of faint emission line sources such as quasars. When a grating is used, spectral lines may be identified according to the formula

$$\lambda = \frac{d}{n} \; \frac{L}{\sqrt{L^2 + D^2}} \tag{2}$$

where D = the distance from grating surface to film plane in millimeters. The head of the comet is the appropriate target for this method, and the dispersion should again be oriented perpendicular to the tail. The field-of-view is that of the telescope. A larger objective makes this method more efficient. For stellar sources, the focal ratio is unimportant, but it does matter for an extended source, in particular, the coma of a comet. When observing extended objects, a fast optical system is advantageous because a shorter exposure time is desirable. Thus, there is a trade-off between wavelength uncertainty and photographic speed, and instrument choice should be based on the goals of the research when possible.

A slitless spectrograph can be constructed using a commercially available visual spectroscope (Lacroix, 1982). Light focused by the telescope objective is collimated by an eyepiece, passes through the direct vision prism system,* and is then focused by a standard camera lens attached to a camera body holding the film. This system has all the elements of an observatory spectrograph except the slit. As with objective prism spectroscopy, the determination of spectral line wavelengths is most easily accomplished with a calibration curve (Schmiedeck, 1979) when a prism or prism system is the disperser.

---

* Any prism or grating can be used. The direct vision prism system allows a simple, straight-through design without the bend in the optical axis that a single prism or grating usually requires.

It is especially important to guide on the comet when doing nonobjective
or slitless spectroscopy. With any of the spectroscopic methods described
in this section, poor guiding in the direction of dispersion will smear the
spectral lines making interpretation difficult or impossible.

If more than one spectroscopic method is used, separate report forms
should be maintained for recording the data. In addition to the instrumental
information requested on the report form, it will be necessary to calibrate
each observer's spectrograph. For objective spectroscopy, a 5-minute trailed
exposure (i.e., turn the clock drive off and let the star trail _perpendicular_
to the dispersion) on one of the stars in Table 4-4, Part II, should be made.
Spectral lines should be clearly visible on the photograph. Calibration
of nonobjective and slitless spectrographs should also use one of the
stars in Table 4-4. The spectrum should clearly show spectral lines and
should be broadened slightly (1 to 2 mm) perpendicular to the dispersion;
this is easily accomplished by varying the telescope's clock drive rate.
Calibration spectra should be included with each roll of film or, better
yet, on each night of comet observation. (The second-to-last paragraph of
the Photography section also discusses calibration and should be read.) A
line on the spectroscopic observations report form should be filled out
for each star observation. Comet spectra should especially be obtained
any time the comet's spectrum and a bright star's spectrum fit in the same
field of view; use two lines on the report form and identify the star by
its right ascension and declination or other designator.

Spectroscopic observers can expect to contribute quantitative photometric data on the wavelengths of each tail ion and in coma ions. Professional
astronomers are not likely to make many observations by the methods described
here. Thus, amateurs have a real opportunity to contribute to the astrophysical study of comets.

Explanation of Spectroscopic Observation Report Form

Camera Lens Focal Length and f/ - Focal length in mm and focal ratio (f/#) of the camera lens used.

Comet or Star Designation - Indicate if the negative has a spectrum of the comet or of a calibration star. If it is a stellar spectrum, give the name or right ascension and declination of the star.

Duration - Give the exposure time in minutes and tenths of minutes for long exposures and in seconds for short exposures.

Faintest Star - Give the magnitude of the faintest star visible to the naked eye (to within 0.5 magnitude) on the star chart in Part II containing the comet's position for the night of observation. M, T, C, or Z should be included with the stellar magnitude when moonlight, twilight, city lights, or zodiacal light (Table 3-1), respectively, interfere with the observations.

Film Name - The manufacturer and film type should be listed as well as the ISO (ASA/DIN). If the film has been hypersensitized, give the method (dry nitrogen, forming gas, silver nitrate rinse, alcohol rinse, etc.), the temperature of the solution, and the duration of the soak. For cooled emulsion photography, give the temperature at which the exposure was made. Indicate the applicable temperature scale. The film processing method should include developer, temperature, and time (or write "commercial" if processed professionally).

Grating - Give the number of grooves per millimeter of the grating and the order or wavelength for which it is blazed.

Guiding - Indicate the guiding method used (see the Photography section for detailed descriptions).

Negative Number - Give the number of the negative to which the exposure details apply.

Notes - Include comments on special circumstances, unusual events, exceptions, and deviations recognized during the observation or in the methods used to make the observation.

Objective, Nonobjective, or Slitless - Check the method of spectroscopy used.

Observer - Each individual observer should use his/her own observing report form, complete with his/her name.

Prism - Give the angle in degrees between the two prism faces used to make the spectrum and the glass type, if known.

Projection Distance - For nonobjective spectroscopy give the distance from the ruled grating surface or center of the prism face to the film.

Site - Give the name of your observing site used for the reported observation. If the site is not one listed on your Observer Index form, include the longitude, latitude, and altitude. Longitude, latitude, and altitude are available on topographic maps available at appropriate government offices and some sporting goods and map stores. If these coordinates are not available, give the nearest town, village, or major landmark and its distance and direction from the site.

Telescope - Give the telescope type, objective diameter, and focal length. This needs to be filled in only if the nonobjective or slitless spectroscopic methods are used.

UT Date - Give the date based on Universal Time.

UT Date Range - Give the first and last UT dates included for the spectrograms described on the report form.

UT Start - Give the Universal Time of the beginning of the observation.

SPECTROSCOPIC OBSERVATION REPORT FORM

(Use separate report forms for different spectroscopic methods.)

UT Date Range ________________________________   Observer ________________________________

Telescope Type _______________   Aperture ____________   Focal Length ___________
    or
Camera Lens Focal Length _____________   f/_____________

Objective _________   Nonobjective _________   Slitless _________

Film Name ____   ISO (ASA/DIN) ________________

Hypersensitized in ________________________________ at ____ °C/°F for ____ hours

Emulsion cooled to ____ °C/°F

Developed in _____________ at ____ °C/°F for ____ minutes

Guiding: Computed ____   Micrometer ____   On Condensation ____

       Tangent X-hairs ____   X-hairs on Coma ____

Grating ________ gr/mm, blaze order _________.
                    Projection Distance _________ mm
Prism Apex Angle ____ °   Glass Type ___________

Exposures

Negative Number	Comet or Star Designation	UT Date	UT Start	Duration	Faintest Star	Site

Notes:

Submit contact prints or duplicate slides <u>with your name on them</u> to the Photography Recorder.

Photoelectric photometry is an area in which a growing number of amateurs participate. Genet (1983) gives a useful introduction to the subject. The techniques of cometary photometry are not substantially different in principle from those of stellar photometry. Details of method are slightly different. Halley's Comet provides an opportunity for amateurs to make contributions in a new area that wasn't available to them even fifteen years ago or to professionals at the last apparition of Halley's Comet. A strong professional effort is planned in this area, but observations by amateurs will help to fill gaps due to bad weather and inadequate coverage in geographic longitude. Amateurs can also provide valuable data because of the smaller image scale of the telescopes they use (if their equipment is sufficiently sensitive to avoid loss of low surface brightness data in equipment noise). This means a much larger area of the coma can be examined photometrically than can be done conveniently with large observatory telescopes.

Amateurs can make several types of observation. Accuracies of ±0.01 magnitudes should be achieved. The determination of coma and central condensation surface brightness at specific wavelengths is the most obvious activity. When made with the proper filters, these observations lead to cometary gas and dust abundances, models of coma chemistry, data on the variation of cometary activity with heliocentric distance, and other results. Studies of short-period time variations are valuable. Coma and tail intensity profiles are of great interest if they are made with sufficient scale and sensitivity. Simple studies of the amount of polarization of light in the coma and tail can be made, but they are difficult to interpret. The photometry of a star seen through the coma or tail has only recently been accomplished for the first time.

The problems of comet photometry fall in two areas. Certain changes in hardware and in observational methods are necessary for the generation of useable data.

The most important hardware change is in the filters. The standard U, B, and V filters used for stellar photometry are virtually useless on comets because their large bandwidths do not allow separation of the contributions of the gas from those of dust in the light of the comet.* Professional astronomers will be using the standard filters listed in Table 9-1 as well as many other narrow passband interference filters for their photometric work on the coma. Attempts are being made to obtain inexpensive substitutes for amateur use, but this may not be possible.

---

* A V magnitude is crudely convertible to a visual magnitude. Studies in the area of photoelectric vs. visual magnitudes are interesting to some researchers, but the value of such data is questionable. U, B, and V filters can be used for transits and occultations of stars by the comet before (mentioned elsewhere in this section). These filters are, as mentioned before, very poor substitutes for a set of comet filters for making measurements of the comet.

TABLE 9-1

Cometary Photometry Filters

Species	Source	Central Wavelength	Bandwidth
Continuum	dust	3650 Å	80 Å
CN	gas	3875	39
C_3	gas	4060	73
CO^+/N_2^+	gas	4260	65
Continuum	dust	4856	85
C_2	gas	5114	90
H_2O^+	gas	7000	175
Continuum	dust	7195	150

The size of the diaphragm used and the image scale of the telescope are directly related. Most large professional telescopes have large image scales and use small diaphragms. An important amateur contribution can be made when the comet head is at its greatest angular extent. The small image scales of amateur telescopes combined with a large diaphragm allow much more of the head to be measured than is easily possible with professional telescopes, since photomultiplier tubes have a limited aperture. The diaphragm size in linear and angular measures must be determined with great accuracy for proper reduction of the data.

In the procedures area, the data are reduced in the usual manner, but more data must be obtained in a more careful manner. Extinction and sky brightness corrections must be made very carefully for each observation because the comet will often be observed low on the horizon around sunrise and sunset. Furthermore, comets are extended objects of very low surface brightness in their outer parts. Average extinction values for the amateur's observatory should never be used. The book by Hall and Genet (1981) is an excellent guide to standard data reduction techniques. The books by Ghedini (1982) and Henden and Kaitchuck (1982) may also prove useful.

Photoelectric magnitudes of the comet should be measured in a manner similar to that for measuring stars. The sequence of measurement should be dark current, comparison star, comet and sky several times, comparison star, dark current, and then the sequence should be repeated. Each sequence should consist of observations with at least one molecule filter and one adjacent continuum filter. After several sequences with one diaphragm, the set of observations should be repeated with a diaphragm of another size. If there is any evidence of clouds or a change in atmospheric transparency (for example, a comparison star has a weaker signal even though it's higher in the sky) photometric observations should be suspended. There may be occasions when conditions are perfect and there simply isn't enough time to get necessary calibration data with the comet data. In such a situation there's no point in making observations.

Standard stars are listed in Tables 4-1, 4-2, and 4-3 in Part II. Sky brightness measurements should be made at least 1° away from the comet head and tail to avoid contamination by the faint outer coma and tail (which will not be visible to the eye), but no more than 5° away. This is crucial to the validity of the comet measurements. Dark current measurements are not necessary in each sequence if past experience has shown it to be stable with time, temperature, etc. (It is not needed in single photomulti- plier tube systems if sky measurements are made carefully.) The most valuable results are possible only with a tube cooled to reduce the dark current.

The observing sequence should be modified to include more comet measure- ments with identical settings if a significant change in apparent brightness is detected. The time of each raw comet head measurement should be recorded to the nearest minute. Knots or disconnected pieces of the ion tail could also be followed through their evolution, but a CO^+ filter is necessary for such observations.

Time studies of intensity variations of the coma or central condensation (large or small diaphragm, respectively) are useful for studies of nuclear activity, rotation, and the comet's interaction with the sun. Gas (the C_3 filter is recommended) or dust filters may be used, and observations should be made continuously through the night. Only one or two filters may be used during such observations.

In any circumstances, it is important to have the same portion of the comet in the diaphragm for each measurement and to record the location in some objective fashion. This requires care and skillful technique.

With a small diaphragm, intensity profiles across the comet can be obtained by allowing the comet to drift through the field of view. The drift should start at least $1/2°$ away from the comet and be at a smooth, constant, known rate (using either the Earth's rotation or the telescope axis drive motors), and the position of the path across the comet should be known. Beware of background stars included in the scan that can falsify the profile. This type of observation can be made in gas and dust filters on the comet. Extinction measurements should be made before and after the drift.

Polarization measurements of the coma are not of obvious value, in general, although there are special cases of transient activity where they would be. They should be made using either a rotating polarizing filter or with several filters oriented in different directions. It is important to know the direction of polarization with respect to north (i.e., the position angle measured from north through east) of each measurement. Prisms or military surplus polarizers should be used, not the camera store variety. (Camera store polarizers are not as efficient as other polarizers.) Again, a clear record of the location for which data are taken is absolutely neces- sary.

Photometry of a star as the nucleus, coma, or tail passes in front of it
(an occultation) would be very interesting. The normal techniques and filters
of stellar photometry should be used, but brightness measurements of the portion of the comet involved in the event without the star (after the occultation) will yield the appropriate intensity value for subtraction. Pure star
readings and dark sky measurements should still be included at least 1° away
from visible cometary features. If an occultation by the nucleus is likely,
accurate timings (based on radio time signals) of the disappearance and reappearance of the star should be made using the filter allowing the highest
signal to noise ratio with the photometer. Continous observations are
necessary for this work and calibration observations should be suspended
for the duration of the possible occultation period.

These projects give amateurs a variety of ways to make potentially
valuable contributions to the research effort on Halley's Comet. A great
deal of care must be taken to obtain useable photometry. Remember that
doing one project continuously and effectively is more valuable than a
variety of projects with intermittent results from each.

Explanation of Photoelectric Photometry Report Form

<u>Air Mass</u> - Give the computed number of air masses through which the observation was made.

<u>Counts</u> - For raw data from a strip chart recorder, this would be the fraction of full scale to three decimals. For reduced data, the results should be in magnitudes or absolute MKS units (Wm^{-2}).

<u>Data</u> - Indicate if the data listed in the table are raw or reduced.

<u>Diaphragm</u> - List the diameter in seconds of arc of the photoelectric photometer diaphragm used.

<u>Filter</u> - Give the name or transmission characteristics of the filter used. For polarizing filters, give the position angle (PA) of the transmitted polarization on the sky. PA is defined to be 0° for due north and increases through 90° due east, 180° due south, and 270° due west. In the field-of-view, with the clock drive off, the last piece of an object drifting out of the field is the eastern piece. Thus, PA is well-defined even in circumpolar regions of the sky.

<u>Notes</u> - Include comments on special circumstances, unusual events, exceptions, and deviations recognized during the observation or in the methods used to make the observation or to do the data reduction.

<u>Object</u> - State whether the data are for the comet, sky, dark current, or a comparison star. For a comparison star, identify it from the list of standard stars.

<u>Observer</u> - Each individual observer should use his/her own observing report form, complete with his/her name.

<u>Photometer</u> - Give the requested information on your photometer.

<u>Portion of Comet Observed</u> - Supply a complete description of the position of the diaphragm on the comet.

<u>Site</u> - Give the name of your observing site used for the reported observation. If the site is not one listed on your Observer Index form, include the longitude, latitude, and altitude. Longitude, latitude, and altitude are available on topographic maps available at appropriate government offices and some sporting goods and map stores. If these coordinates are not available give the nearest town, village, or major landmark and its distance and direction <u>from</u> the site.

<u>Telescope</u> - Give the telescope type, objective diameter, and focal length.

<u>UT Date</u> - Give the date based on Universal Time.

<u>UT Start - End</u> - Give the Universal Time of the start and end of the observation.

9-5

PHOTOELECTRIC PHOTOMETRY REPORT FORM

UT Date ______________________________   Observer ____

Site ________________________________

Telescope Type __________________   Aperture __________ Focal Length __________

Photometer: Detector Type ______________________________   Cooled ___   Uncooled ___

Amplifier Type ______________________________

Recording System: Analog ______   Digital______

Detector Voltage ____________   Amplifier Gain ____________

N.D. Filter Used On Stellar Standards? Yes ___   No ___

Portion of Comet Observed ____

____

Data are Raw ___   Reduced ___

Object	Diaphragm	Filter[1]	UT Start - End	Air Mass	Counts[2]

Notes

---

[1] Attach a copy of transmission curves for any nonstandard filters used. This only needs to be done once, when the first report of its use is submitted.

[2] For raw data from a strip chart recorder, this would be the fraction of full scale to three decimals. For reduced data, the results should be in magnitudes or absolute MKS units (Wm^{-2}).

10. METEOR OBSERVATIONS

Meteor studies are an area not discussed in detail in IHW Science Working Group documents. Because of the established relationship of comets and meteor streams, studies of the meteor showers that may be associated with Halley's Comet may yield useful data applicable to the understanding of the comet. The data required are easily obtained and provide both the novice and the experienced observer an opportunity to contribute to an area in which few professional astronomers participate.

Cook (1973) gives data on the two meteor streams which may be associated with Halley's Comet: the η Aquarid meteors in spring and Orionid meteors in fall. These meteor showers occur when the Earth makes its closest approaches to Halley's orbit. The meteoroids released by the comet are slowly spread out along the whole cometary orbit and also out of the plane of the orbit. The meteors we see at Earth's closest approach to the orbit may be caused by particles released at a Halley apparition centuries ago which have since been perturbed into orbits that intersect with the Earth's orbit.

The η Aquarid meteors apparently radiate from right ascension (α) 22^h 22^m and declination (δ) $-1°54'$ (1950) in Aquarius on the night of maximum rate. They are identifiable from April 21 to May 12.* The maximum rate occurs approximately May 3 (Millman 1980); other sources suggest May 4 (McKinley 1961) and May 5 (Lovell, 1954; Hughes, 1978). In past years, the rate for single observers has been about 20 meteors per hour, and the shower rate is above five meteors per hour for three days centered on maximum. No major variations in rate have been observed in past years. These meteors are among the fastest shower meteors, with measured velocities of 65.5 km/s.

The Orionids radiate from several radiants in the vicinity of α 6^h 18^m and δ $15°$ $48'$. The extreme dates of identification are from October 2 to November 7. The maximum rate is about 25 meteors per hour on October 21 and is above six per hour for two days. Increased rates were observed in the 1922 shower. Measured Orionid velocities at 66.4 km/s are even faster than η Aquarid velocities. Persistent trains are more common than with other meteor showers.

Equatorial observers are favored by the low declinations of the radiants, but in terms of meteor rates observers in both temperate zones will not be greatly affected by the lower altitudes of the radiants. Observers at high latitudes will find that these shower radiants rise shortly before sunrise, thus limiting their counting period when the radiant is above the horizon.

---

* A variety of references were searched to find the longest period of visibility for shower members, which are the dates given here. Other references cite much shorter periods. There is no question that the count rate more than 2-3 days away from maximum is much lower than the rate near maximum. Observations made off-maximum are valuable because relatively few people make them, and occasionally the maximum rate will occur off the predicted time of maximum.

Because much is already known about these meteor showers, IHW meteor observers are asked to make contributions mainly in selected areas of meteor astronomy. These areas were chosen with the expectation that they will yield significant data relating to the showers and the comet.

Visual counts and spectrophotography of meteors are the two topics IHW collaborators should concentrate on. Photographs, especially those obtained from two stations for triangulation (height determination) purposes and those obtained with a chopper (high-speed, rotating shutter) for velocity determination are also of great interest but are extremely difficult to obtain. Methods of obtaining height and velocity data from photographs will be found in some of the books in Appendix II. The American Meteor Society has kindly offered to advise and assist serious efforts to obtain two station photographs or spectrograms (discussed more below) of Halley-related meteors obtained during the course of observations.

Visual counts are simple to make. The unit of counting is the number of shower members and non-shower meteors (counted separately) seen by one observer in one hour. Individuals observing together should keep <u>separate</u> tallies, and each should face different areas of the sky. Data reduction is simplified if the starting time is on the hour: Universal Time should be used.

Meteor observers should be in a reclining, comfortable position with their eyes naturally falling about 50° above the horizon. The method of counting must be included in the personal equation for each observer. Pencil and paper can be used but require the observer to take his/her eyes off the sky. Gate counters or finger-keyed supermarket hand adding machines are better used for counting shower and nonshower meteors. An observer who desires to contribute more can use a tape recorder with microphone on/off switch to record data including shower membership, magnitude, color, whether or not a wake or persistent train was seen, and other data like duration, path length, and elevation.* The tapes can later be transcribed onto the standard report form.

Qualitative impressions or quantitative data on the showers are also requested. These could take the form of comments like "large number of fragmenting meteors" or "37% ± 3% of observed meteors were deep red."

In the unlikely event of a meteor storm, the time interval for counts or sky area observed may have to be reduced to a known value. A note explaining what procedural changes were made should be included on the report form.

---

* Such detailed observations require a great deal of practice to decrease the psychological biases inherent in visual observations. Regular observations of the major and minor meteor showers visible throughout the year should be made to maintain a high degree of skill for these observations. Many meteor organizations are pleased to receive year-round observations.

Counts of telescopic meteors made simultaneously with those obtained
without optical aid can also be useful. The telescope aperture, field-of-
view, magnification, and approximate right ascension and declination of the
center of the field of view should be reported with the number of meteors
and hours of observing.

Meteor spectra can be photographed with an objective grating or prism
placed in front of a fast camera lens (see the section on Spectroscopic
Observations in this manual). High speed films should be used, but most
hypersensitizing techniques will not improve the chances of catching a
meteor. A clock drive should be used to track the sky, and the camera
should be oriented so that the dispersion is perpendicular to the line from
the aim point to the radiant (i.e., perpendicular to the path a shower
meteor will follow). The aim point should be about 40° from the radiant,
but this is by no means a hard and fast rule.

Capturing meteors on film is very dependent on good luck, fast film
speed, and the use of enough film to allow the good luck to occur. Exposures should be short to avoid undue sky fog (this requires experimentation
at the observation site). If the observer suspects the capture of a meteor
spectrum, the exposure should be ended immediately, and a star calibration
spectrum should be obtained as described in the Spectroscopic Observations
section.

Under certain circumstances it is possible to observe VHF radio signals
scattered from meteor trails. The American Meteor Society has an observation program which uses low power aeronautical beacons for all-weather
meteor counts. Amateur radio operators and other communications monitors,
especially those in the equatorial zone, are encouraged to participate in a
special study of the η Aquarid and Orionid meteors. Details will be sent to
potential observers who indicate an interest in this in a letter to the
Coordinator for Amateur Observations (address on the Observer Index form).

To get good statistics on the η Aquarid and Orionid meteor showers, it
is important to get as many observations as possible. Meteor observations
in the years before Halley's return provide a baseline for comparison with
the showers during and after the comet's appearance. It is important to
begin these meteor studies as soon as possible and to continue them indefin-
itely.

As with all other observations, practice improves the quality of the
observations through training and familiarization. Regular observations
also allow the observer to perceive changes as they occur. Observers are
encouraged to begin and then continue studies of the η Aquarid and Orionid
meteor showers and also to monitor the other active meteor showers visible
during the year. The American Meteor Society and other organizations are
happy to receive visual observations of all major and minor meteor showers.

Explanation of Meteor Observation Report Forms

Camera Lens Focal Length and f/ - Focal length in mm and focal ratio (f/#) of the camera lens used.

Cloud Cover - Sketch in the approximate amount of cloud cover in each octant of the sky. Note the cloud type or thickness.

Count Method - Indicate the method used for keeping the meteor count.

Dark Adaptation Time - State how long your eyes have had to dark adapt before beginning your visual observations. (A minimum of 20 minutes is necessary.)

Duration - Give the exposure time in minutes and tenths of minutes for long exposures and in seconds for short exposures.

Facing Direction - Give the approximate direction faced (e.g., NW, SSE, WSW, S, etc.)

Faintest Star - Give the magnitude of the faintest star visible to the naked eye (to within 0.5 magnitude) in the center of the meteor field of view at the beginning and end of each meteor count period. M, T, C, or Z should be included with the stellar magnitude when moonlight, twilight, city lights, or zodiacal light (Table 3-1), respectively, interfere with the observation.

Film Name - The manufacturer and film type should be listed as well as the ISO (ASA/DIN). The film processing method should include developer, temperature (indicate the scale used), and time (or write "commercial" if processed professionally).

Grating - Give the number of grooves per millimeter of the grating and the order or wavelength for which it is blazed.

Group Observation - Check Yes or No. If Yes, include the names of other observers in the group in the "Notes" section. Meteor counts must be reported individually and on an hourly basis.

Meteor or Star Designation - Indicate if the negative has a spectrum of an η Aquarid or Orionid meteor or of a calibration star. Spectra of nonshower meteors may also be submitted and should be designated "nonshower." If the the spectrum is from a star give the name or right ascension and declination of the star.

Negative Number - Give the number of the negative to which the exposure details apply.

Notes - Include comments on special circumstances, unusual events, exceptions, and deviations recognized during the observation or in the methods used to make the observation.

Number of Meteors - Give the number of shower meteors and non-shower meteors observed during the period of observation.

Observer - Each individual observer should use his/her own observing report form, complete with his/her name.

Prism - Give the angle between the two prism faces used to make the spectrum and the glass type, if known.

Rotating Shutter Chop Frequency; Other Chopper Information - Give the number of breaks per second made by the chopper; give any other relevant details regarding the chopper.

Site - Give the name of your observing site used for the reported observation. If the site is not one listed on your Observer Index form, include the longitude, latitude, and altitude. Longitude, latitude, and altitude are available on topographic maps available at appropriate government offices and some sporting goods and map stores. If these coordinates are not available give the nearest town, village, or major landmark and its distance and direction from the site.

Triangulation: Second Observer, Second Site, Paired Negatives - Give the other observer's name, the other site's name and geographic location, and the number of the pairs of negatives containing the same meteor from the two sites.

UT Date - Give the date based on Universal Time.

UT Date Range - Give the first and last UT dates included for the spectro-grams described on the report form.

UT Start - End - Give the Universal Time of the start/end of the observa-tion.

UT Start - Give the Universal Time of the beginning of the observation.

Viewing Area of Sky - Indicate unlimited or give the size of the area being concentrated on.

VISUAL METEOR OBSERVATION REPORT FORM

UT Date _________________   Observer _____________________

Dark Adaptation Time ___________   Site _________________________

Cloud Cover   Count Method: Written _____________________

                                Counter _____________________

                                Tape Recorder ______________

Facing Direction ____________   Group Observation? Yes ___ No ___

Viewing Area of Sky: Unrestricted ___ Limited to ___° x ___°

UT Start - End	Faintest Star	Number of Meteors	
		Shower	Non-shower

Notes:

METEOR PHOTOGRAPHY INFORMATION REPORT FORM

UT Date Range _____________   Observer _____________________

Camera Lens Focal Length _____________ f/_____________

Film Name _______________________ ISO (ASA/DIN) _____________________

Developed in ____________ at ____ °C °F for ____ minutes

Grating _____ gr/mm, blaze order ______.

Prism apex angle ____ ° Glass Type _________________

Rotating Shutter Chop Frequency _______ . Other Chopper Info.: _____________

____

Exposures

Negative Number	Meteor or Star Designation	UT Date	UT Start	Duration	Faintest Star	Site

Triangulation: Second Observer _______________________

Second Site ____

Paired Negative Numbers _______________________________

____

(A separate report form should be completed for the second site.)

Notes:

Submit contact prints or duplicate slides <u>with your name on them</u> to the Meteor Recorder.

APPENDIX A

Addresses of Organizations and Publications

AAS Photobulletin, Robert J. Leacock, Subscription Manager, 211 Space Sciences Building, University of Florida, Gainesville, FL 32611, USA.

AMERICAN METEOR SOCIETY, Dept. of Physics and Astronomy, State University of New York at Geneseo, Geneseo, NY 14454, USA.

AMERICAN ASSOCIATION OF VARIABLE STAR OBSERVERS (AAVSO), 187 Concord Ave., Cambridge, MA 02138, USA.

ASSOCIATION OF LUNAR AND PLANETARY OBSERVERS (ALPO), Journal of the ALPO (The Strolling Astronomer), P. O. Box 3AZ, University Park, NM 88003, USA ; COMET SECTION: Dennis Milon, 8 Grant St., Maynard, MA 01754, USA.

Astronomical Almanac, U. S. Naval Observatory, 34th and Massachusetts Ave. N.W., Washington, DC 20390, USA.

ASTRONOMICAL LEAGUE, Significant Event Announcements, Don Archer, Executive Secretary, P. O. Box 12821, Tucson, AZ, 85732, USA.

ASTRONOMICAL SOCIETY OF THE PACIFIC (ASP), Mercury, 1290 24th Ave., San Francisco, CA 94122, USA.

Astronomy, Astromedia Corp., 625 E. St. Paul Ave., P. O. Box 92788, Milwaukee, WI 53202, USA.

AUSTRALIAN COMET SECTION, David Seargent, 156 Entrance Rd., The Entrance, NSW 2261, AUSTRALIA.

BAA COMET SECTION, Michael J. Hendrie, "Overbury," 33 Lexden Rd., West Bergholt, Colchester, Essex CO6 3BX, GREAT BRITAIN.

BAA Handbook, BRITISH ASTRONOMICAL ASSOCIATION, Burlington House, Piccadilly, London, W1V ONL, UNITED KINGDOM.

BAA METEOR SECTION, George Spalding, 2 Hyde Rd., Denchworth, Wantage, Oxon OX12 ODR, UNITED KINGDOM.

John Bortle, W. R. Brooks Observatory Circulars, W. R. Brooks Observatory, Gold Rd., Stormville, NY 12582, USA.

BRITISH METEOR SOCIETY, Robert A. Mackenzie, 26 Adrian St., Dover, Kent, CT17 9AT, UNITED KINGDOM.

CENTRAL BUREAU FOR ASTRONOMICAL TELEGRAMS, Smithsonian Astrophysical Observatory, Cambridge, MA 02138, USA.

COMET AND MINOR PLANET SECTION, RASNZ, Alan C. Gilmore, Mt. John University Observatory, P. O. Box 20, Lake Tekapo, South Canterbury, NEW ZEALAND.

Comet News Service, McDonnell Planetarium, 5100 Clayton Rd., St. Louis, MO 63110, USA.

A-1

(Dutch Comet Section) WERKGROEP KOMETEN, Dr. Reinder J. Bouma, Bekemaheerd, 9737
PR Groningen, NETHERLANDS.

FRENCH CNRS COMET COORDINATING GROUP, c/o M. C. Festou, Laboratoire d' Aeronomie
CNRS, Reduit de Verrieres, 91370 Vierrieres-le-Buisson, FRANCE.

Graphic Timetable of the Heavens, Scientia, Inc., 1815 Landrake Rd., Baltimore,
MD 21204, USA.

HUNGARIAN COMET OBSERVING NETWORK, Tuboly Vince, H-9915 Hegyhatsal, F6 U't 19.
SZ, HUNGARY.

INTERNATIONAL AMATEUR-PROFESSIONAL PHOTOELECTRIC PHOTOMETRY Association
(IAPPP), c/o R. C. Wolpert, Belmont Observatory, 144 Neptune Avenue, North
Babylon, NY 11704, USA.

International Comet Quarterly, Daniel Green, Smithsonian Astrophysical Observatory, 60 Garden St., Cambridge, MA 02138, USA. (Do not use ICQ anywhere
in the address.)

INTERNATIONAL HALLEY WATCH, Jet Propulsion Laboratory, T-1166, 4800 Oak Grove
Drive, Pasadena, CA, 91109, USA.

(Japanese Amateur Comet Observers), HOSHINO HIROBA, Akira Kamo, 5-10 Shimazaki-
cho, Wakayama-shi 640, JAPAN.

Japan Astronomical Circular, Takeshi Urata, 271-2, Umegaya, Shimizu, Shizuoka,
JAPAN 424.

JAPAN ASTRONOMICAL STUDY ASSOCIATION, Keiichi Saijo, National Science Museum,
Ueno Park, Taito, Tokyo, JAPAN 110.

Meteor News, c/o Wanda Simmons, Route 3, Box 424-99, Callahan, FL 32011, USA.

METEOR SECTION, RASNZ, Ken Morse, P. O. Box 2241, Wellington, NEW ZEALAND.

NIPPON METEOR SOCIETY, Yasuo Yabu, 878 Maruyama, Omihachiman, Shiga, JAPAN 523.

ORIENTAL ASTRONOMICAL ASSOCIATION, Ichiro Hasegawa, 2-3-11, Saidaiji-Nogamicho,
Nara, JAPAN 631.

RASC Observer's Handbook, ROYAL ASTRONOMICAL SOCIETY OF CANADA, 124 Merton St.,
Toronto, Ontario, M4S, 2Z2, CANADA.

Sky and Telescope and Sky Publications, Sky Publishing Corp., 49 Bay State Rd.,
Cambridge, MA 02238, USA.

Sky-Gazers Almanac or similar diagram in the January Sky and Telescope each year.
The address of Sky and Telescope is given above.

APPENDIX B

Bibliography

Bobrovnikoff, N. T., 1931, "Halley's Comet in Its Apparition of 1909-1911," Publications of the Lick Observatory, Univ. of California Press, Vol. 17 Pt. II, p. 305.

Bortle, J.E., 1980, "How to Observe Comets" in the "Sky and Telescope" Guide to the Heavens, L. J. Robinson, ed. (also published in Sky and Telescope, Mar. 1981, p. 210), Cambridge, MA., Sky Publishing Corp.

Brandt, J. C., 1981, Comets--Readings from Scientific American, San Francisco, W. H. Freeman & Co.

Brandt, J. C. and R. D. Chapman, 1981, Introduction to Comets, Cambridge, Cambridge Univ. Press.

Brandt, J. C., B. Donn, J. M. Greenberg, and J. Rahe, eds., 1981, Modern Observational Techniques for Comets, Pasadena, NASA-JPL Publication 81-68.

Brandt, J. C., L. D. Friedman, R. L. Newburn, and D. K. Yeomans, eds., 1980, The International Halley Watch--Report of the Science Working Group, Pasadena, NASA - JPL Publication TM 82181 (400-88).

Brown, P. L., 1973, Comets, Meteorites and Men, New York, Taplinger Publishing Co., 1973.

Calder, N., 1981, The Comet is Coming, New York, Viking Press.

Cook, A. F., 1973, "A Working List of Meteor Streams," Evolutionary and Physical Properties of Meteoroids, Washington, DC, NASA, SP-319.

"Designations of Magnitude References," International Comet Quarterly, Dept. of Physics and Astronomy, Appalachian State Univ., Boone, NC, 1981 Apr., p. 47.

de Vaucouleurs, G., 1961, Astronomical Photography, London, Faber and Faber.

Edberg, S. J., 1982 Mar., "Exploring the Stars with a Spectrograph" in "Observers Page," Sky and Telescope, p. 311.

Eicher, D.J., 1983 Feb., "Capturing Deep Sky Objects on Paper" in "Gazer's Gazette," Astronomy, p. 35.

Everhart, E., 1982 Sep., "Constructing a Measuring Engine" in "Gleanings for ATM's," Sky and Telescope, p. 279.

Feijth, H., 1980 Oct., "Of Sequences and Comparison Star Magnitudes" in International Comet Quarterly, Dept. of Physics and Astronomy, Appalachian State Univ., Boone, NC, p. 73.

Genet, R.M., 1983 Feb., "Backyard Photoelectric Photometry," in "Equipment Atlas," Astronomy, p. 51.

Ghedini,S., 1982, <u>Software for Photometric Astronomy</u>, Richmond, VA, Willmann-Bell.

Hall, D. S. and R. M. Genet, 1981, <u>Photoelectric Photometry of Variable Stars</u>, Fairborn, OH, Fairborn Observatory.

"Halley's Spin Period is 10^{hr} 19^{min}," <u>Comet News Service</u>, McDonnell Planetarium, St. Louis, No. 80-2, 1980 Apr. 24.

Henden, A. and R. Kaitchuck, 1982, <u>Astronomical Photometry</u>, New York, Van Nostrand Reinhold.

Herald, D., 1981 Apr., "Visual Magnitudes and the SAO Catalog," <u>International Comet Quarterly</u>, Dept. of Physics and Astronomy, Appalachian State Univ., Boone, NC, p. 43.

Hughes, D. W., 1978, "Meteors," <u>Cosmic Dust</u>, J. A. M. McDonnell, ed., New York, John Wiley & Sons.

Huling, John Jr., 1981 July, a letter to the editor in "Letters," <u>Sky and Telescope</u>, p. 16.

Kimball, D. S., 1957 May, "Visual Aurora Observing During the IGY," <u>Sky and Telescope</u>, p. 327.

King, E. S., 1931, <u>A Manual of Celestial Photography</u>, Boston, Eastern Science Supply Co.

Lacroix, D. P., 1982 July, "Stellar Spectra the Easy Way" in "Observer's Page," <u>Sky and Telescope</u>, p. 99.

Lines, R. D., 1973a June, "A Simple Micrometer Microscope for Off-Axis Guiding in Comet Photography," <u>The Strolling Astronomer, Journal of the Association of Lunar and Planetary Observers</u>, Las Cruces, p.97.

Lines, R. D., 1973b June, "How to Use a Micrometer Microscope to Guide for Comet Photography," <u>The Strolling Astronomer, Journal of the Association of Lunar and Planetary Observers</u>, Las Cruces, p. 98.

Lovell, A. C. B., 1954, <u>Meteor Astronomy</u>, London, Oxford Univ. Press.

Majden, E. P., 1978 June, "Conventional Meteor Spectroscopy for Amateurs," <u>Meteor News</u>, Wanda Simmons, Callahan Astronomical Society, Callahan, Florida, No. 41, p. 1.

Marcus, J. N., 1981 Jan. 12, "Observing Comets for Nuclear Rotation," <u>Comet News Service</u>, McDonnell Planetarium, St. Louis, No. 81-1.

Marsden, B. G., 1979, "Comet Halley and History," <u>Space Missions to Comets</u>, NASA CP-2089, Washington, DC.

Marsden, B. G., 1982 Sep., "How to Reduce Plate Measurements" in "Gleanings for ATM's," <u>Sky and Telescope</u>, p. 284.

Marsden, B. G., 1983, Catalog of Cometary Orbits, Hillside, NJ, Enslow.

Mayall, R. N. and M. W. Mayall, 1968, Skyshooting - Photography for Amateur Astronomers, New York, Dover Publications, Inc.

McKinley, D. W. R., 1961, Meteor Science and Engineering, New York, McGraw-Hill Book Co., Inc.

McLeod, Norman W., III, Visual Meteor Observing - Notes and Instructions, American Meteor Society, State Univ. of New York, Geneseo.

Middlehurst, B.M. and G.P. Kuiper, eds., 1963, The Moon, Meteorites, and Comets, Chicago, Univ. of Chicago Press.

Miller, W. C., 1980, "Dark Adaptation: Its Nature and Preservation," AAS Photobulletin, 1980 No. 2 (Consecutive No. 24), p. 18, Working Group on Photographic Materials, AAS, Rochester, Kodak.

Millman, P. M., 1957 May, "An IGY Program of Meteor Observing for Amateurs," Sky and Telescope, p. 317.

Millman, P. M., 1980, "Meteors, Fireballs and Meteorites," Observer's Handbook 1981, Percy, J. R., ed., Toronto, Univ. of Toronto.

Morris, C. S., 1979 Apr. 27, "A New Method for Estimating Cometary Brightness," Comet News Service, McDonnell Planetarium, St. Louis, No. 79-1.

Morris, C. S., 1980 Oct., "A Review of Visual Comet Observing Techniques, I," International Comet Quarterly, Dept. of Physics and Astronomy, Appalachian, State Univ., Boone, NC, p. 69.

Morris, C. S., 1981a Jan., "A Review of Visual Comet Observing Techniques, II," International Comet Quarterly, Dept. of Physics and Astronomy, Appalachian State Univ., Boone, NC, p.3.

Morris, C. W., 1981b July, "A Review of Visual Comet Observing Techniques, III," International Comet Quarterly, Dept. of Physics and Astronomy, Appalachian State Univ., Boone, NC, p.89.

Morris, C. S. and D. W. E. Green, 1982 June, "The Light Curve of Periodic Comet Halley 1910 II," Astronomical Journal, Vol. 87 (No. 6), p. 918.

Newburn, R. L., Jr., J. F. Bell, and T. B. McCord, 1981 Mar.,"Interference Filter Photometry of Periodic Comet Ashbrook-Jackson," Astronomical Journal Vol. 86 (No.3), p.469.

Newburn, R. L., Jr., and D. K. Yeomans, 1982, "Halley's Comet", in Ann. Rev. of Earth and Planetary Sciences, Vol. 10, Palo Alto, Annual Reviews, Inc.

Nye, R. A., 1981 Nov., "Arizona Amateur's Photoelectric Photometer" in "Gleanings for ATM's," Sky and Telescope, p. 496.

Patterson, G. N., 1981, Handbook of Astrophotography for Amateur Astronomers, Saskatoon, RASC Saskatoon Centre.

Patterson, J. and P. Michaud, 1980 Feb., "Photographing Stellar Spectra" in "Photography in Astronomy," _Astronomy_, p. 39.

Paul, H. E., 1960, _Outer Space Photography for the Amateur_, New York, Amphoto.

Penhallow, W. S., 1978 Nov., "Notes on a 16-Inch Astrometric Reflector" in "Gleanings for ATM's," _Sky and Telescope_, p. 455.

Perrine, C. D., 1934, "Observaciones del Cometa Halley Durante su Aparicion 1910," _Pub. Resultados del Observatorio Nacional_, Argentino, Vol. 25.

Polman, J., 1977 May, "A Homemade Filar Micrometer" in "Gleanings for ATM's," _Sky and Telescope_, p. 391.

Povenmire, H. R., 1980, _Fireballs, Meteors, and Meteorites_, Indian Harbor Beach, FL, JSB Enterprises, Inc.

"Projects for May with a Sky Crossbow" in "Celestial Calendar," _Sky and Telescope_, 1981 May, p. 417.

Rahe, J., B. Donn, and K. Wurm, 1969, _Atlas of Cometary Forms_, Washington, DC, NASA, SP-198.

Richardson, R. S., 1967, _Getting Acquainted with Comets_, New York, McGraw-Hill Book Co.

Roach, F. E. and P. M. Jamnick, 1958 Feb., "The Sky and Eye," _Sky and Telescope_, p. 164.

Roach, F. E. and G. Van Biesbroeck, 1954 Mar., "The Zodiacal Light and the Solar Corona," _Sky and Telescope_, p. 144.

Roques, P.E. and P.K. Whitt, 1971 Feb., "Dust Between the Planets," _Griffith Observer_, Griffith Observatory, Los Angeles, p. 22.

Roth, G. D., ed., 1975, _Astronomy: A Handbook_, Cambridge, MA., Sky Publishing Corp.

Schmiedeck, W., 1979 Apr., "A Simple Technique for Recording the Sun's Spectrum," in "Gleanings for ATM's," _Sky and Telescope_, p. 395.

Sherrod, P. C. with T. L. Koed, 1981, _A Complete Manual of Amateur Astronomy_, Englewood Cliffs, NY, Prentice-Hall, Inc.

Sherrod, C., _Observations of Meteors_, Little Rock, The Astronomical Unit.

Sherrod, C., _Studies of the Solar System Part I--Comets_, Little Rock, The Astronomical Unit.

Sidgwick, J. B., 1980, _Amateur Astronomer's Handbook_, 4th edition, Hillside, NJ, Enslow.

Sidgwick, J. B., 1982, _Observational Astronomy for Amateurs_, 4th edition, Hillside, NJ, Enslow.

Star Catalog, Washington, DC, Smithsonian Institution (Smithsonian Astro-
     physical Observatory), 1966.

Struve, O., 1956 Sep., "The Spectra of Comets," Sky and Telescope, p. 489.

Tatum J., 1982a Apr., "The Measurement of Comet Positions," Journal of the
     Royal Astronomical Society of Canada, Vol. 76, No. 2, p. 97.

Tatum J., 1982b June, "The Calculation of Comet Ephemerides," Journal of the
     Royal Astronomical Society of Canada, Vol. 76, No. 3, p. 157.

Waber, R. and R. McPherson, 1967 May, "Photographing Star Spectra" in
     "Observer's Page," Sky and Telescope, p. 322.

Wilkening, L.W., ed., 1982, Comets, Tucson, University of Arizona Press.

Wood, F. B., ed., 1963, Photoelectric Astronomy for Amateurs, New York,
     Macmillan Co.

Worley, C. E., 1961 Sep., "The Construction of a Filar Micrometer," Sky and
     Telescope, p. 140.

Yeomans, D. K,, 1977, "Comet Halley - the Orbital Motion," Astronomical
     Journal, Vol. 82, p. 435.

Yeomans, D. K., 1981, The Comet Halley Handbook, Pasadena, NASA - JPL
     Publication 400-91.

Yeomans, D. K. and T. Kiang, 1981, "The Long Term Motion of Comet Halley,"
     Monthly Notices of the Royal Astronomical Society, Vol. 197, p. 633.

International Halley Watch Amateur Observers' Manual for Scientific Comet Studies

Part II. Ephemeris and Star Charts

Stephen J. Edberg

The research described in this manual was performed at the Jet Propulsion Laboratory, California Institute of Technology, under contract with the National Aeronautics and Space Administration.

ABSTRACT

This manual describes the International Halley Watch, comets and observing techniques, and provides information on periodic Comet Halley's apparition for its 1986 perihelion passage. Part I gives detailed instructions for observation projects valuable to the International Halley Watch in six areas of study: (1) visual observations, (2) photography, (3) astrometry, (4) spectroscopic observations, (5) photoelectric photometry, and (6) meteor observations. Part II includes an ephemeris for Comet Halley for the period 1985-1987 and star charts showing its position from November 1985 through May 1986.

FOREWORD

This manual has been written for the advanced amateur astronomer.
Part I provides instructions on the proper methods of generating meaningful
scientific data on comets. The novice can learn general observing techniques
while learning the methods in this manual. Part II contains an ephemeris
and star charts for finding the comet and making observations of it.

This manual does not teach basic observing, telescope adjustment, or
data reduction techniques. There are many books available for such purposes.
It should be stressed that the most important single thing an amateur can do
to advance his skills is to use them. Practice provides part of the training
necessary to advance in the techniques of skillful, scientific observing.
Both novice and experienced observers should observe comets as they appear
in preparation for Comet Halley's apparition in 1985-86.

It is a sad fact that many professional astronomers are unaware of the
careful, professional-quality work which amateur astronomers are capable of
and have, in some cases, been doing for years. There is now a movement to
call people who practice astronomy without pay "nonprofessionals" in an
effort to improve the image of the hardworking, dedicated amateur who does
reputable research. While "amateur" and "nonprofessionals" are both accurate,
the latter is less fluent and has not been used in the text. It is my hope
that activities like the IHW Amateur Observation Net will demonstrate to
professionals that their unpaid fellow astronomers--amateurs--are worthy of
the respect sought with the "nonprofessional" noun.

Amateurs have made and can continue to make important contributions
to cometary research. Dedication to the effort is all that's required.

                                                          S. E.

IHW AMATEUR OBSERVER'S MANUAL

Part II - Ephemeris and Charts

CONTENTS

1. REPORT FORM SAMPLES AND GLOSSARY OF TERMS

Report Form Samples

The following sample forms are included in Part II:

- Observer Index

- Visual Observations

- Drawing

- Photography

- Astrometry

- Spectroscopic Observations

- Photoelectric Photometry

- Visual Meteor Observations

- Meteor Photography

OBSERVER INDEX

Please tear out this form and fill it in as completely as possible if you plan to submit observations to the IHW. Also fill in the duplicate in Part II for your own records. It is important to read Sections 2 and 4 and the section describing your area of participation in Part I of this manual before submitting this index form. Return this form to Stephen Edberg (Jet Propulsion Laboratory, 4800 Oak Grove Dr., T-1166, Pasadena, California 91109, USA).

Name (Last, First) __________________________________   Telephone:

Mailing Address _________________________________   Day ____________________
                                                            area code + number

____

____   Night ____________________
                                                              area code + number

____

Areas of Participation: (check all that apply)

    ____ Visual Observations   ____ Spectroscopic Observations
    ____ Photography   ____ Photoelectric Photometry
    ____ Astrometry   ____ Meteor Studies

List Regular Observation Site(s). Longitude, latitude, and altitude may be determined using topographic maps.

Name	Longitude	Latitude	Altitude
1. __________________________	__________	__________	__________
2. __________________________	__________	__________	__________
3. __________________________	__________	__________	__________
4. __________________________	__________	__________	__________

Provide the information requested on telescopes you expect to use including the units of measurement. Indicate the site numbers (from the list above) where the telescope has a permanent mount or where a portable mount is regularly used for visual (V), photographic (PG), and/or photoelectric (PE) observing. Binoculars users should state the power and aperture (e.g., 7x50) with the word binoculars under telescope type and skip the next two columns. Meteor observers should write meteor and visual, photographic, or radio under telescope type and give the site numbers where these observations are usually made.

Telescope Type	Aperture	Focal Length	Mounting Site # Perm.	Port.	Observing V	PG	PE
__________	______	______	______	______	__	__	__
__________	______	______	______	______	__	__	__
__________	______	______	______	______	__	__	__
__________	______	______	______	______	__	__	__

List equipment planned for use in photography <u>not already listed</u> as a telescope.
This can include Schmidt or aerial cameras or interchangeable lenses belonging
to your photographic system.

Camera   Focal Length   f/ratio   Notes

__________   __________   ______   ________________________

__________   __________   ______   ________________________

__________   __________   ______   ________________________

__________   __________   ______   ________________________

<u>Photometric Equipment:</u>

Photomultiplier Tube ___________________________   Cooled ____   Uncooled ____

Electronics: Photon Counting ___________   Analog ________

<u>Miscellaneous Accessories:</u>

Diffraction Grating Source or Manufacturer   ______________________________

________ gr/mm, blaze order _______

Prism: Glass Type ______   Apex Angle ________

Rotating Meteor Shutter Chop Rate ___________

I understand that the data I contribute to the International Halley Watch may
be used by IHW Archive users and that my contribution will be acknowledged in
the usual manner in any publications resulting from such use. I understand
that I may also publish my data in any manner I choose.

____
                    Signature   Date

                    Novice   Moderate   Expert
Level of Observational Experience   _____   _______   _____
General Astronomical Observations   _____   _______   _____
Comet Observations   _____   _______   _____
Meteor Observations   _____   _______   _____

Are you planning on traveling to the southern hemisphere to observe Halley's
Comet in March or April 1986? Yes ___   No___

Additional Comments:

VISUAL OBSERVATION REPORT FORM

Observer _______________________

| UT Date and Time | M. M. | Coma (Total) Magnitude | Chart No. | Instrument | | | Magnification | Coma Dia. | D.C. | Tail Length | PA | Faintest Star | Dark Adapted | Site | Notes |
				Aperture	Type	f/									

DRAWING INFORMATION REPORT FORM

UT Date _______________________________   Observer ________________________________

Faintest Star ____________________________   Site ___________________________________

Instrument Aperture ___________________   Type _______________   f/__________

Seeing ____

UT Start ________________________________   UT End ________________________________

Magnification(s) Used ____

Filter(s) Used ____

Features   Type   ID#   PA

                _________________   _________________   _________________

                _________________   _________________   _________________

                _________________   _________________   _________________

                _________________   _________________   _________________

                _________________   _________________   _________________

                _________________   _________________   _________________

                _________________   _________________   _________________

Indicate the orientation (north and east) in the drawing and the scale (minutes of arc per millimeter).

Notes:

PHOTOGRAPHIC INFORMATION REPORT FORM

UT Date Range ___________________________ Observer ___________________________

Instrument Focal Length _______________ f/ _______________

Photographic Method: PF __   NP __   EP __   A __   EFL = _______ mm

Film Name ___________________________________ ISO (ASA/DIN) _______________

Hypersensitized in _______________________________ at ____ °C/°F for ____ hours

Emulsion cooled to ____ °C/°F

Developed in _______________________________ at ____ °C/°F for ____ minutes

Guiding: Computed ___   Micrometer ___   On Condensation ___

          Tangent X-hairs ___   X-hairs on Coma ___

Exposures

Negative Number	UT Date	UT Start	Duration	Filter	Faintest Star	Site

Notes:

Submit contact prints or duplicate slides <u>with your name on them</u> to the
Photography Recorder.

ASTROMETRIC DATA REPORT FORM

UT Date Range _________________________ Observer _______________________________

Instrument Focal Length _______________ f/ ______________

Photographic Method: PF ___   NP ___   EP ___   A ___   EFL = _____ mm

Film Name _________________________________ ISO (ASA/DIN) _________________

Hypersensitized in _____________________________ at ____ $^{\circ}$F for ____ hours

Emulsion cooled to ____ $^{\circ}$F

Developed in _______________________________ at ____ $^{\circ}$F for ____ minutes

Guiding: Computed ___   Micrometer ___   On Condensation ___

          Tangent X-hairs ___   X-hairs on Coma ___

Exposures

Nega-tive #	UT Date	UT of Mid-Exposure	Dura-tion	Site	Computed α	Computed δ	Comet Image

Notes:

Submit this form to the Coordinator for Amateur Observations.

SPECTROSCOPIC OBSERVATION REPORT FORM

(Use separate report forms for different spectroscopic methods.)

UT Date Range ______________________________ Observer ____________________________

Telescope Type ________________ Aperture ____________ Focal Length __________
      or
Camera Lens Focal Length ______________ f/______________

Objective _________ Nonobjective _________ Slitless _________

Film Name ____ ISO (ASA/DIN) ______________

Hypersensitized in _________________________________ at ____ °C/°F for ____ hours

Emulsion cooled to ____ °C/°F

Developed in ______________ at ____ °C/°F for ___ minutes

Guiding: Computed ____ Micrometer ____ On Condensation ____

        Tangent X-hairs ____ X-hairs on Coma ____

Grating ________ gr/mm, blaze order ________.

Prism Apex Angle ____ ° Glass Type __________   Projection Distance ________ mm

Exposures

Negative Number	Comet or Star Designation	UT Date	UT Start	Duration	Faintest Star	Site

Notes:

Submit contact prints or duplicate slides <u>with your name on them</u> to the Photography Recorder.

PHOTOELECTRIC PHOTOMETRY REPORT FORM

UT Date _____________________   Observer _________________________________

Site _______________________________

Telescope Type ________________   Aperture _________ Focal Length _________

Photometer: Detector Type ____________________   Cooled ___   Uncooled ___

             Amplifier Type _______________________

             Recording System: Analog ______   Digital______

             Detector Voltage ___________   Amplifier Gain ___________

             N.D. Filter Used On Stellar Standards? Yes ___   No ___

Portion of Comet Observed ____

____

Data are Raw ___   Reduced ___

Object	Diaphragm	Filter[1]	UT Start – End	Air Mass	Counts[2]

Notes

---

[1] Attach a copy of transmission curves for any nonstandard filters used. This only needs to be done once, when the first report of its use is submitted.

[2] For raw data from a strip chart recorder, this would be the fraction of full scale to three decimals. For reduced data, the results should be in magnitudes or absolute MKS units (Wm^{-2}).

VISUAL METEOR OBSERVATION REPORT FORM

UT Date _____________________   Observer _______________________________

Dark Adaptation Time _____________   Site ___________________________________

Cloud Cover   Count Method: Written __________________

                                                        Counter __________________

                                                        Tape Recorder _____________

Facing Direction _____________   Group Observation? Yes ___ No ___

Viewing Area of Sky: Unrestricted ___ Limited to ___° x ___°

UT		Faintest Star	Number of Meteors	
Start	– End		Shower	Non-shower
	//////// ////////			
//////// ////////				
	//////// ////////			
//////// ////////				
	//////// ////////			
//////// ////////				
	//////// ////////			
//////// ////////				
	//////// ////////			
//////// ////////				

Notes:

METEOR PHOTOGRAPHY INFORMATION REPORT FORM

UT Date Range _____________   Observer ___________________________

Camera Lens Focal Length _____________ f/_____________

Film Name _______________________ ISO (ASA/DIN) _____________________

Developed in ___________ at ____ °C/°F for ____ minutes

Grating _____ gr/mm, blaze order ______.

Prism apex angle ____ ° Glass Type __________________

Rotating Shutter Chop Frequency _______ . Other Chopper Info.: _____________

____

Exposures

Negative Number	Meteor or Star Designation	UT Date	UT Start	Duration	Faintest Star	Site

Triangulation: Second Observer ___________________________

Second Site ____

Paired Negative Numbers ______________________________________

____

(A separate report form should be completed for the second site.)

Notes:

Submit contact prints or duplicate slides with your name on them to the
Meteor Recorder.

Comprehensive Report Form Explanation

<u>Air Mass</u> - Give the computed number of air masses through which the observation was made.

<u>Camera Lens Focal Length and f/</u> - Focal length in mm and focal ratio (f/#) of the camera lens used.

<u>Chart No.</u> - The number of the IHW amateur manual's chart (<u>not page number</u>) used for comparison stars.

<u>Cloud Cover</u> - Sketch in the approximate amount of cloud cover in each octant of the sky. Note the cloud type or thickness.

<u>Coma Dia.</u> - The coma diameter observed in minutes of arc. Give the long and short dimensions of an elliptical coma.

<u>Coma (Total) Magnitude</u> - The coma's estimated magnitude; should be reported to the nearest 0.1 magnitude. If the magnitude is based on stars whose magnitudes are underlined, underline your estimate.

<u>Comet Image</u> - Note whether or not the measured comet image was diffuse or had an obvious central condensation. The discussion of degree of condensation in the Visual Observations section may be instructive.

<u>Comet or Star Designation</u> - Indicate if the negative has a spectrum of the comet or of a calibration star. If it is a stellar spectrum, give the name or right ascension and declination of the star.

<u>Count Method</u> - Indicate the method used for keeping the meteor count.

<u>Counts</u> - For raw data from a strip chart recorder, this would be the fraction of full scale to three decimals. For reduced data, the results should be in magnitudes or absolute MKS units (Wm^{-2}).

<u>Dark Adaptation Time</u> - State how long your eyes have had to dark adapt before beginning your visual observations. (A minimum of 20 minutes is necessary.)

<u>Dark Adapted</u> - Indicate Y (yes) or N (no) if you were dark adapted when the comet observation was made.

<u>Data</u> - Indicate if the data listed in the table are raw or reduced.

<u>D.C.</u> - Give the degree of condensation of the coma.

<u>Diaphragm</u> - List the diameter in seconds of arc of the photoelectric photometer diaphragm used.

<u>Duration</u> - Give the exposure time in minutes and tenths of minutes for long exposures and in seconds for short exposures.

<u>EFL</u> - Give the effective focal length of the photographic method used.

<u>Facing Direction</u> - Give the approximate direction faced (e.g., NW, SSE, WSW, S, etc.)

1-13

Faintest Star - Give the magnitude of the faintest star visible to the
naked eye (to within 0.5 magnitude) on the star chart in Part II containing the comet's position for the night of observation. For meteor observations give the magnitude of the faintest star visible to the naked eye
(to within 0.5 magnitude) in the center of the meteor field of view at
the beginning and end of each meteor count period. M, T, C, or Z should
be included with the stellar magnitude when moonlight, twilight, city
lights, or zodiacal light (Table 3-1), respectively, interfere with the
observation.

Film Name - The manufacturer and film type should be listed as well as the
ISO (ASA/DIN). If the film has been hypersensitized, give the method (dry
nitrogen, forming gas, silver nitrate rinse, alcohol rinse, etc.), the
temperature of the solution, and the duration of the soak. For cooled
emulsion photography, give the temperature at which the exposure was made.
Indicate the applicable temperature scale. The film processing method
should include developer, temperature, and time (or write "commercial" if
processed professionally).

Filter - Give the name or transmission characteristics of the filter used or
list the color and designation of the filter. Examples are Yellow, Wratten
12; Light Blue, Schott BG 38; etc. For Schott, Corning, and similar filters,
include the thickness of the glass. For polarizing filters, give the position
angle (PA) of the transmitted polarization on the sky. PA is defined to be
0° for due north and increases through 90° due east, 180° due south, and 270°
due west. In the field of view, with the clock drive off, the last piece of
an object drifting out of the field is the eastern piece. Thus, PA is well-
defined even in circumpolar regions of the sky.

Filter(s) Used - List the filters used when the drawing was made.

Grating - Give the number of grooves per millimeter of the grating and the
order or wavelength for which it is blazed.

Group Observation - Check Yes or No. If Yes, include the names of other
observers in the group in the "Notes" section. Meteor counts must be
reported individually and on an hourly basis.

Guiding - Indicate the guiding method used (see the Photography section for
detailed descriptions).

Instrument - For Aperture, give the objective diameter in centimeters. Type
describes the optical system (refractor, Newtonian, Cassegrain, Schmidt-
Cassegrain, binoculars, etc.), and f/ is the focal ratio of the instrument.

Instrument Focal Length and f/ - Focal length in mm and focal ratio (f/#)
of the instrument used.

Magnification - Found by dividing the telescope focal length by the focal
length of the eyepiece used for the observation.

Magnification(s) Used - List the magnification(s) used when the drawing
was made.

<u>Meteor or Star Designation</u> - Indicate if the negative has a spectrum of an η Aquarid or Orionid meteor or of a calibration star. Spectra of nonshower meteors may also be submitted and should be designated "nonshower." If the spectrum is from a star give the name or right ascension and declination of the star.

<u>M.M.</u> - The magnitude estimation method used:

      B = Bobrovnikoff, S = Sidgwick, M = Morris

<u>Negative Number</u> - Give the number of the negative to which the exposure details apply.

<u>Notes</u> - Include further explanation for the Comet Image classification, if necessary, and comments on special circumstances, unusual events, exceptions, and deviations recognized during the observation or in the methods used to make the observation or to do the data reduction.

<u>Number of Meteors</u> - Give the number of shower meteors and nonshower meteors observed during the period of observation.

<u>Object</u> - State whether the data are for the comet, sky, dark current, or a comparison star. For a comparison star, identify it from the list of standard stars.

<u>Objective, Nonobjective, or Slitless</u> - Check the method of spectroscopy used.

<u>Observed α, Observed δ</u> - Give the right ascension (α) to two decimal places in seconds of time and declination (δ) to one decimal place in seconds of arc computed for the comet's position for the UT date and time of mid-exposure.

<u>Observer</u> - Each individual observer should use his/her own observing report form, complete with his/her name.

<u>PA</u> - Position angle of tail(s). For a curved tail give the distance from the nucleus where the measurement applies. Give the method used for determining it (plot, clock face, calibrated eyepiece). PA is defined to be 0° for due north and increases through 90° due east, 180° due south, and 270° due west. In the field of view, with the clock drive off, the last piece of an object drifting out of the field is the eastern piece. Thus, PA is well-defined even in circumpolar regions of the sky.

<u>Photometer</u> - Give the requested information on your photometer.

<u>Photographic Method</u> - Indicate the type based on the following light paths:

      Principal Focus (PF): telescope objective - film
      Negative Projection (NP): telescope objective - negative lens - film
      Eyepiece Projection (EP): telescope objective - eyepiece - film
      Afocal (A): telescope objective - eyepiece - camera lens - film

<u>Portion of Comet Observed</u> - Supply a complete description of the position of the diaphragm on the comet.

1-15

Prism - Give the angle in degrees between the two prism faces used to make the spectrum and the glass type, if known.

Projection Distance - For nonobjective spectroscopy give the distance from the ruled grating surface or prism face to the film.

Rotating Shutter Chop Frequency; Other Chopper Information - Give the number of breaks per second made by the chopper; give any other relevant details regarding the chopper.

Seeing - Estimation of the seeing quality in seconds of arc or describe the seeing in some other standard manner.

Site - Give the name of your observing site used for the reported obervation. If the site is not one listed on your Observer Index form, include the longitude, latitude, and altitude. Longitude, latitude and altitude are available on topographic maps available at appropriate government offices and some sporting goods and map stores. If these coordinates are not available give the nearest town, village, or major landmark and its distance and direction from the site.

Tail Length - Reported in degrees and tenths of degrees. Use two lines if two tails are visible.

Telescope - Give the telescope type, objective diameter, and focal length. For spectroscopic observations this needs to be filled in only if the nonobjective or slitless spectroscopic methods are used.

Triangulation: Second Observer, Second Site, Paired Negatives - Give the other observer's name, the other site's name and geographic location, and the number of the pairs of negatives containing the same meteor from the two sites.

UT Date - Give the date based on Universal Time.

UT Date and Time - Local dates and times should be converted to UT as explained in the Universal Time subsection, p. 4-6. Times given in UT should be accurate to ± 5 minutes. A decimal date (e.g., Nov. 12, 12:00 UT = Nov. 12.50 UT) can be included if desired. Decimal dates should be accurate to ±0.005 day.

UT Date Range - Give the first and last UT dates included for the photographs or spectrograms described on the report form.

UT of Mid-Exposure - Give the Universal Time of the middle of the photographic exposure, accurate to 1 second.

UT Start - End - Give the Universal Time of the start/end of the observation.

Viewing Area of Sky - Indicate unlimited or give the size of the area being concentrated on.

2. SELECTED EPHEMERIS DATA

FROM

<u>The Comet Halley Handbook</u>
(second edition)

BY

DONALD K. YEOMANS

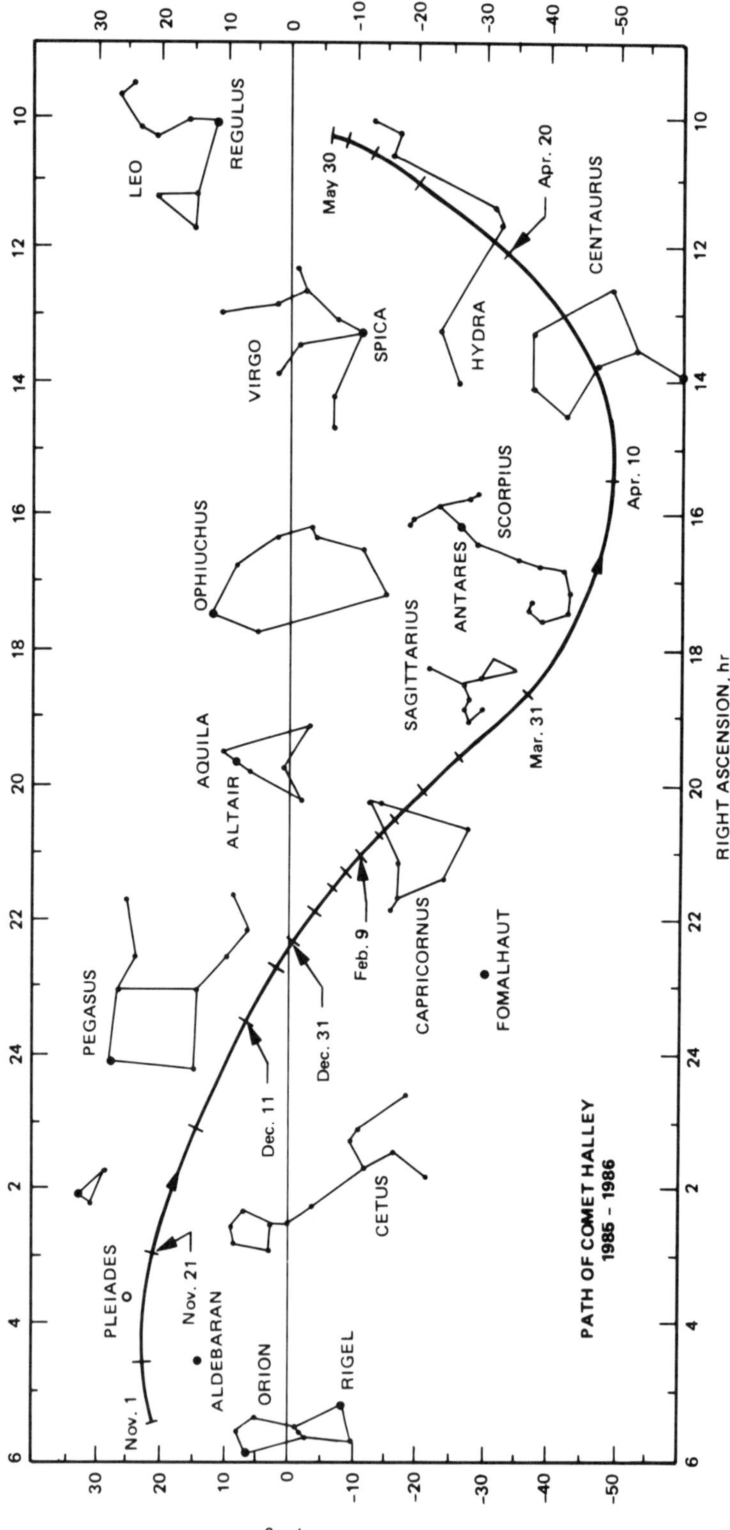

Path of Comet Halley on the Celestial Sphere During November 1985-May 1986

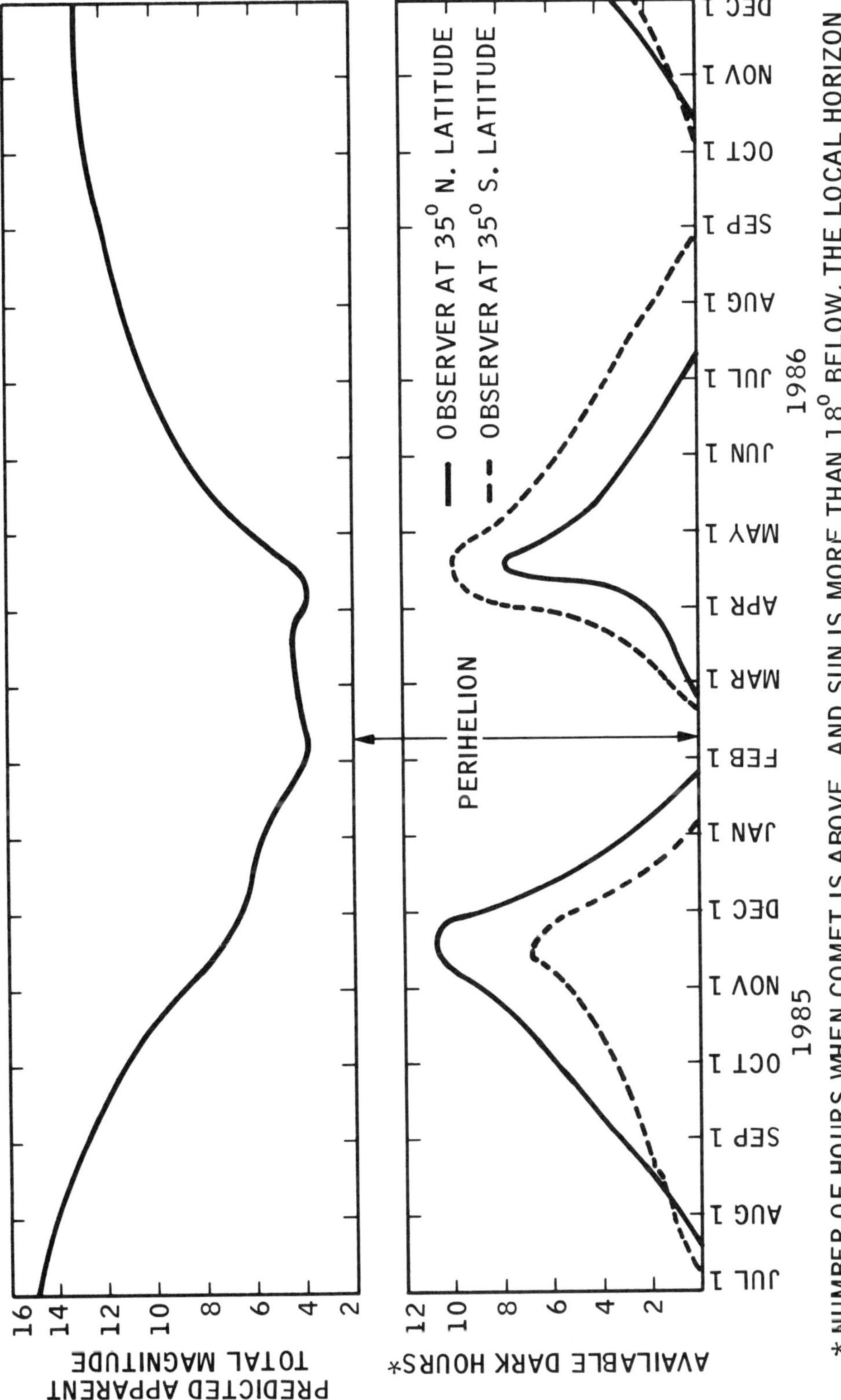

Comet Halley 1985-1986 Ground-Based Observing Conditions

Ground-Based Observing Data, COMET HALLEY 1985-1986

Date (1985)		North 45°	Lat. 30°	South 30°	Lat. 45°	Apparent Magnitudes M_1	Date (1986)		North 45°	Lat. 30°	South 30°	Lat. 45°	Apparent Magnitudes M_1
Jan.	1	11.6	10.9	6.8	3.5	16.7	Jan.	6	2.6	2.3	0.5	0	5.6
	11	10.7	10.0	6.9	3.9	16.6		16	1.3	1.1	0	0	5.1
	21	9.7	9.1	7.2	4.6	16.5		26	0	0	0	0	4.4
	31	8.7	8.1	5.6	5.3	16.4	Feb.	5	0	0	0	0	3.9
Feb.	10	7.7	7.2	5.0	3.7	16.4		15	0	0	0	0	4.1
	20	6.8	6.4	4.4	3.3	16.3		25	0	0.3	0.7	0.5	4.3
Mar.	2	5.8	5.5	3.9	2.9	16.2	Mar.	7	0.2	0.9	2.0	2.0'	4.5
	12	4.9	4.7	3.4	2.5	16.2		17	0.5	1.5	3.3	3.7	4.5
	22	4.0	4.0	2.9	2.1	16.1		27	0.7	2.3	5.3	6.2	4.3
Apr.	1	3.2	3.2	2.4	1.8	16.0	Apr.	6	0	3.8	9.1	9.4	4.0
	11	2.3	2.5	1.9	1.4	15.9		16	6.0	8.3	10.0	9.9	4.4
	21	1.4	1.7	1.5	1.0	15.9		26	6.2	8.0	9.2	10.0	5.5
May	1	0.5	1.0	1.0	0.6	15.7	May	6	5.5	5.1	7.7	8.3	6.6
	11	0	0.3	0.5	0.2	15.6		16	2.7	4.3	6.7	7.3	7.5
	21	0	0	0	0	15.5		26	1.9	3.5	6.0	6.5	8.3
	31	0	0	0	0	15.3	Jun.	5	1.0	2.8	5.3	5.8	8.9
Jun.	10	0	0	0	0	15.1		15	0.2	2.1	4.6	5.1	9.4
	20	0	0	0	0	14.9		25	0	1.5	4.0	4.5	9.9
	30	0	0	0	0	14.7	Jul.	5	0	0.9	3.4	3.9	10.3
Jul.	10	0	0	0.3	0.1	14.5		15	0	0.4	2.7	3.2	10.7
	20	0	0.5	0.8	0.5	14.2		25	0	0	2.1	2.5	11.0
	30	0.5	1.2	1.3	0.9	13.9	Aug.	4	0	0	1.5	1.9	11.3
Aug.	9	1.4	1.9	1.7	1.3	13.6		14	0	0	0.9	1.2	11.6
	19	2.3	2.6	2.1	1.6	13.2		24	0	0	0.3	0.5	11.8
	29	3.2	3.3	2.5	1.9	12.8	Sep.	3	0	0	0	0	12.0
Sep.	8	4.1	4.0	2.9	2.2	12.4		13	0	0	0	0	12.2
	18	5.0	4.8	3.3	2.5	11.9		23	0	0	0	0	12.3
	28	5.9	5.6	3.8	2.8	11.3	Oct.	3	0	0	0	0	12.5
Oct.	8	6.9	6.4	4.3	3.1	10.7		13	0	0.1	0.4	0	12.6
	18	8.0	7.3	4.9	3.5	10.0		23	0.4	0.8	0.8	0.4	12.7
	28	9.2	8.5	5.7	4.1	9.1	Nov.	2	1.2	1.5	1.2	0.7	12.8
Nov.	7	10.7	10.0	6.8	5.0	8.2		12	2.0	2.2	1.7	1.0	12.9
	17	11.1	10.6	7.3	4.9	7.2		22	2.7	2.8	2.3	1.4	13.0
	27	10.6	9.8	7.0	4.2	6.5							
Dec.	7	7.7	7.1	4.4	3.6	6.3							
	17	5.5	5.1	2.8	1.0	6.2							
	27	3.9	3.6	1.5	0	6.0							

Note: (1) For a particular observer's latitude, the number of dark hours is defined as the time interval during which the Sun is below the local horizon by at least 18° and the comet is simultaneously above the local horizon.

(2) Magnitude estimates are based upon the comet's observed behavior in 1909–1910. Predictions are for ideal observing conditions.

For charts of the comet's path above the horizon as seen from different latitudes, turn to Section 5.

Explanation of Symbols for Ephemeris

J.D. = Julian Date (Ephemeris Time)

R.A. 1950.0 DEC. = Geocentric right ascension and declination referred to the mean equator and equinox of 1950.0. A light time correction has been applied.

R.A. APPN DEC. = Apparent geocentric right ascension and declination. Light time, annual aberration, and nutation corrections have been applied, and R.A. and Dec. have been precessed to the ephemeris date.

DELTA = Geocentric distance of comet in AU.

DELDOT = Geocentric velocity of comet in km/sec.

R = Heliocentric distance of comet in AU.

RDOT = Heliocentric velocity of comet in km/sec.

M_1 = Total magnitude = 5.47 + 5.0 log (DELTA) + 11.1 log (R), preperihelion. Postperihelion, the corresponding equation is total magnitude = 4.94 + 5.0 log (DELTA) + 7.68 log (R).

M_2 = Nuclear magnitude = 14.1 + 5.0 log (DELTA) + 5.0 log (R).

NOTE: In cases where M_1 is not computed, the corresponding column is filled with zeros (.0)

THETA = Sun-Earth-Comet angle in degrees.

BETA = Sun-Comet-Earth angle in degrees.

MOON = Comet-Earth-Moon angle in degrees.

NOTES: 1. The following osculating orbital elements are consistent with the following ephermeris:

Epoch	2446480.50000	1986 FEB. 19.00000 (E.T.)
Perihelion Passage	2446470.95174	1986 FEB. 9.45174 (E.T.)
Perihelion Distance in AU	0.5871047	
Eccentricity	0.9672760	
Argument of Perihelion	111.84809	
Longitude of Ascending Node	58.14536	
Inclination	162.23928	

2. Angles are in degrees and are referred to the ecliptic and equinox of 1950.0.

3. The style II nongravitational parameters are:

$$A_1 = (0.0565 \pm 0.0213) \times 10^{-8} \text{ AU/(day)}^2$$

$$A_2 = (0.0154 \pm 0.0001) \times 10^{-8} \text{ AU/(day)}^2$$

YR	MN	DY	HR	J.D.	R.A. (1950)	DEC.	R.A. APPN	DEC.	DELTA	DELDOT	R	RDOT	TMAG	NMAG	THETA	BETA	MOON
1985	6	4	.0	2446220.5	5 15.148 +17	25.97	5 17.168 +17	28.27	4.76	-14.89	3.76	-18.72	15.2	20.4	8.5	2.3	170
1985	6	5	.0	2446221.5	5 15.736 +17	27.98	5 17.757 +17	30.25	4.75	-15.36	3.75	-18.74	15.2	20.3	7.9	2.1	158
1985	6	6	.0	2446222.5	5 16.328 +17	29.97	5 18.349 +17	32.21	4.74	-15.83	3.73	-18.77	15.2	20.3	7.3	2.0	145
1985	6	7	.0	2446223.5	5 16.924 +17	31.94	5 18.946 +17	34.15	4.73	-16.31	3.72	-18.79	15.2	20.3	6.8	1.9	132
1985	6	8	.0	2446224.5	5 17.523 +17	33.89	5 19.546 +17	36.08	4.72	-16.79	3.71	-18.82	15.2	20.3	6.3	1.7	119
1985	6	9	.0	2446225.5	5 18.126 +17	35.83	5 20.150 +17	37.99	4.71	-17.26	3.70	-18.84	15.1	20.3	6.0	1.6	107
1985	6	10	.0	2446226.5	5 18.732 +17	37.75	5 20.757 +17	39.88	4.70	-17.74	3.69	-18.87	15.1	20.3	5.7	1.6	95
1985	6	11	.0	2446227.5	5 19.342 +17	39.66	5 21.367 +17	41.75	4.69	-18.22	3.68	-18.90	15.1	20.3	5.5	1.5	83
1985	6	12	.0	2446228.5	5 19.956 +17	41.54	5 21.981 +17	43.61	4.68	-18.70	3.67	-18.92	15.1	20.3	5.4	1.5	71
1985	6	13	.0	2446229.5	5 20.572 +17	43.41	5 22.598 +17	45.44	4.67	-19.17	3.66	-18.95	15.1	20.3	5.5	1.5	60
1985	6	14	.0	2446230.5	5 21.192 +17	45.26	5 23.219 +17	47.26	4.66	-19.65	3.65	-18.97	15.0	20.2	5.6	1.6	48
1985	6	15	.0	2446231.5	5 21.814 +17	47.09	5 23.842 +17	49.06	4.65	-20.13	3.64	-19.00	15.0	20.2	5.9	1.6	37
1985	6	16	.0	2446232.5	5 22.440 +17	48.90	5 24.468 +17	50.84	4.63	-20.61	3.63	-19.03	15.0	20.2	6.3	1.8	26
1985	6	17	.0	2446233.5	5 23.068 +17	50.69	5 25.097 +17	52.61	4.62	-21.09	3.61	-19.05	15.0	20.2	6.7	1.9	15
1985	6	18	.0	2446234.5	5 23.698 +17	52.47	5 25.729 +17	54.35	4.61	-21.56	3.60	-19.08	15.0	20.2	7.2	2.0	8
1985	6	19	.0	2446235.5	5 24.331 +17	54.23	5 26.363 +17	56.08	4.60	-22.04	3.59	-19.11	14.9	20.2	7.8	2.2	14
1985	6	20	.0	2446236.5	5 24.967 +17	55.97	5 26.999 +17	57.79	4.58	-22.52	3.58	-19.13	14.9	20.2	8.3	2.4	25
1985	6	21	.0	2446237.5	5 25.605 +17	57.69	5 27.638 +17	59.48	4.57	-22.99	3.57	-19.16	14.9	20.2	9.0	2.5	37
1985	6	22	.0	2446238.5	5 26.244 +17	59.39	5 28.278 +18	1.15	4.56	-23.47	3.56	-19.19	14.9	20.1	9.6	2.7	50
1985	6	23	.0	2446239.5	5 26.886 +18	1.07	5 28.921 +18	2.80	4.54	-23.94	3.55	-19.22	14.9	20.1	10.3	2.9	63
1985	6	24	.0	2446240.5	5 27.530 +18	2.74	5 29.565 +18	4.43	4.53	-24.41	3.54	-19.24	14.8	20.1	11.0	3.1	76
1985	6	25	.0	2446241.5	5 28.175 +18	4.38	5 30.211 +18	6.04	4.51	-24.88	3.53	-19.27	14.8	20.1	11.7	3.3	90
1985	6	26	.0	2446242.5	5 28.822 +18	6.01	5 30.858 +18	7.64	4.50	-25.35	3.51	-19.30	14.8	20.1	12.4	3.6	103
1985	6	27	.0	2446243.5	5 29.470 +18	7.61	5 31.507 +18	9.22	4.49	-25.81	3.50	-19.33	14.8	20.1	13.1	3.8	117
1985	6	28	.0	2446244.5	5 30.120 +18	9.20	5 32.158 +18	10.77	4.47	-26.28	3.49	-19.35	14.8	20.1	13.8	4.0	131
1985	6	29	.0	2446245.5	5 30.771 +18	10.77	5 32.809 +18	12.31	4.46	-26.74	3.48	-19.38	14.7	20.1	14.6	4.2	145
1985	6	30	.0	2446246.5	5 31.423 +18	12.33	5 33.462 +18	13.83	4.44	-27.20	3.47	-19.41	14.7	20.0	15.3	4.4	159
1985	7	1	.0	2446247.5	5 32.076 +18	13.86	5 34.116 +18	15.33	4.42	-27.66	3.46	-19.44	14.7	20.0	16.1	4.7	170
1985	7	2	.0	2446248.5	5 32.730 +18	15.38	5 34.771 +18	16.81	4.41	-28.12	3.45	-19.47	14.7	20.0	16.8	4.9	167
1985	7	3	.0	2446249.5	5 33.385 +18	16.87	5 35.427 +18	18.28	4.39	-28.57	3.44	-19.50	14.6	20.0	17.6	5.1	155
1985	7	4	.0	2446250.5	5 34.040 +18	18.35	5 36.083 +18	19.73	4.37	-29.03	3.42	-19.53	14.6	20.0	18.3	5.4	142
1985	7	5	.0	2446251.5	5 34.697 +18	19.81	5 36.741 +18	21.15	4.36	-29.48	3.41	-19.55	14.6	20.0	19.1	5.6	129
1985	7	6	.0	2446252.5	5 35.353 +18	21.26	5 37.398 +18	22.57	4.34	-29.94	3.40	-19.58	14.6	19.9	19.9	5.8	116
1985	7	7	.0	2446253.5	5 36.011 +18	22.68	5 38.056 +18	23.96	4.32	-30.39	3.39	-19.61	14.5	19.9	20.6	6.1	103
1985	7	8	.0	2446254.5	5 36.669 +18	24.09	5 38.715 +18	25.33	4.31	-30.84	3.38	-19.64	14.5	19.9	21.4	6.3	91
1985	7	9	.0	2446255.5	5 37.326 +18	25.48	5 39.373 +18	26.69	4.29	-31.29	3.37	-19.67	14.5	19.9	22.2	6.5	79
1985	7	10	.0	2446256.5	5 37.984 +18	26.85	5 40.032 +18	28.03	4.27	-31.73	3.36	-19.70	14.5	19.9	22.9	6.8	68
1985	7	11	.0	2446257.5	5 38.642 +18	28.20	5 40.690 +18	29.35	4.25	-32.18	3.35	-19.73	14.4	19.9	23.7	7.0	56
1985	7	12	.0	2446258.5	5 39.300 +18	29.54	5 41.349 +18	30.66	4.23	-32.62	3.33	-19.76	14.4	19.8	24.5	7.3	45
1985	7	13	.0	2446259.5	5 39.957 +18	30.86	5 42.007 +18	31.94	4.21	-33.07	3.32	-19.79	14.4	19.8	25.3	7.5	33
1985	7	14	.0	2446260.5	5 40.614 +18	32.16	5 42.664 +18	33.21	4.19	-33.51	3.31	-19.82	14.4	19.8	26.0	7.7	22
1985	7	15	.0	2446261.5	5 41.270 +18	33.45	5 43.321 +18	34.46	4.17	-33.94	3.30	-19.85	14.3	19.8	26.8	8.0	12
1985	7	16	.0	2446262.5	5 41.926 +18	34.71	5 43.978 +18	35.69	4.15	-34.38	3.29	-19.88	14.3	19.8	27.6	8.2	8
1985	7	17	.0	2446263.5	5 42.580 +18	35.96	5 44.633 +18	36.91	4.13	-34.82	3.28	-19.91	14.3	19.8	28.4	8.5	17
1985	7	18	.0	2446264.5	5 43.234 +18	37.20	5 45.287 +18	38.11	4.11	-35.25	3.27	-19.95	14.2	19.7	29.2	8.7	29
1985	7	19	.0	2446265.5	5 43.886 +18	38.41	5 45.940 +18	39.30	4.09	-35.68	3.25	-19.98	14.2	19.7	30.0	9.0	42
1985	7	20	.0	2446266.5	5 44.537 +18	39.61	5 46.592 +18	40.47	4.07	-36.10	3.24	-20.01	14.2	19.7	30.7	9.2	55
1985	7	21	.0	2446267.5	5 45.186 +18	40.80	5 47.241 +18	41.62	4.05	-36.52	3.23	-20.04	14.2	19.7	31.5	9.5	69
1985	7	22	.0	2446268.5	5 45.833 +18	41.96	5 47.889 +18	42.76	4.03	-36.94	3.22	-20.07	14.1	19.7	32.3	9.7	82
1985	7	23	.0	2446269.5	5 46.479 +18	43.12	5 48.535 +18	43.88	4.01	-37.36	3.21	-20.10	14.1	19.6	33.1	10.0	96

YR	MN	DY	HR	J.D.	R.A. (1950) .0	DEC.	R.A. APPN	DEC.	DELTA	DELDOT	R	RDOT	TMAG	NMAG	THETA	BETA	MOON
1985	7	24	.0	2446270.5	5 47.122	+18 44.25	5 49.179	+18 44.98	3.99	-37.77	3.20	-20.13	14.1	19.6	33.9	10.2	110
1985	7	25	.0	2446271.5	5 47.763	+18 45.37	5 49.821	+18 46.07	3.97	-38.18	3.18	-20.17	14.0	19.6	34.7	10.5	124
1985	7	26	.0	2446272.5	5 48.402	+18 46.48	5 50.461	+18 47.14	3.94	-38.59	3.17	-20.20	14.0	19.6	35.5	10.7	137
1985	7	27	.0	2446273.5	5 49.038	+18 47.57	5 51.098	+18 48.20	3.92	-38.99	3.16	-20.23	14.0	19.5	36.3	11.0	151
1985	7	28	.0	2446274.5	5 49.672	+18 48.65	5 51.732	+18 49.25	3.90	-39.39	3.15	-20.26	14.0	19.5	37.1	11.2	164
1985	7	29	.0	2446275.5	5 50.303	+18 49.72	5 52.364	+18 50.28	3.88	-39.79	3.14	-20.30	13.9	19.5	37.9	11.5	171
1985	7	30	.0	2446276.5	5 50.930	+18 50.77	5 52.992	+18 51.30	3.85	-40.18	3.13	-20.33	13.9	19.5	38.7	11.7	163
1985	7	31	.0	2446277.5	5 51.555	+18 51.81	5 53.618	+18 52.31	3.83	-40.57	3.11	-20.36	13.9	19.5	39.5	12.0	151
1985	8	1	.0	2446278.5	5 52.176	+18 52.83	5 54.240	+18 53.31	3.81	-40.95	3.10	-20.40	13.8	19.5	40.3	12.2	138
1985	8	2	.0	2446279.5	5 52.794	+18 53.85	5 54.859	+18 54.29	3.78	-41.34	3.09	-20.43	13.8	19.4	41.1	12.5	125
1985	8	3	.0	2446280.5	5 53.409	+18 54.85	5 55.474	+18 55.26	3.76	-41.72	3.08	-20.46	13.8	19.4	41.9	12.7	112
1985	8	4	.0	2446281.5	5 54.019	+18 55.84	5 56.085	+18 56.22	3.73	-42.10	3.07	-20.50	13.7	19.4	42.7	13.0	100
1985	8	5	.0	2446282.5	5 54.626	+18 56.82	5 56.692	+18 57.17	3.71	-42.47	3.05	-20.53	13.7	19.4	43.5	13.2	88
1985	8	6	.0	2446283.5	5 55.228	+18 57.79	5 57.295	+18 58.11	3.69	-42.85	3.04	-20.57	13.7	19.3	44.3	13.5	76
1985	8	7	.0	2446284.5	5 55.826	+18 58.75	5 57.893	+18 59.04	3.66	-43.22	3.03	-20.60	13.6	19.3	45.1	13.7	64
1985	8	8	.0	2446285.5	5 56.420	+18 59.70	5 58.487	+18 59.96	3.64	-43.58	3.02	-20.63	13.6	19.3	45.9	14.0	53
1985	8	9	.0	2446286.5	5 57.009	+19 .64	5 59.076	+19 .87	3.61	-43.95	3.01	-20.67	13.6	19.3	46.7	14.2	41
1985	8	10	.0	2446287.5	5 57.592	+19 1.57	5 59.661	+19 1.77	3.58	-44.31	3.00	-20.70	13.5	19.3	47.6	14.5	30
1985	8	11	.0	2446288.5	5 58.170	+19 2.49	6 .239	+19 2.66	3.56	-44.66	2.98	-20.74	13.5	19.2	48.4	14.7	19
1985	8	12	.0	2446289.5	5 58.742	+19 3.40	6 .813	+19 3.55	3.53	-45.02	2.97	-20.77	13.5	19.2	49.2	15.0	9
1985	8	13	.0	2446290.5	5 59.309	+19 4.31	6 1.380	+19 4.42	3.51	-45.36	2.96	-20.81	13.4	19.2	50.0	15.2	10
1985	8	14	.0	2446291.5	5 59.869	+19 5.21	6 1.941	+19 5.30	3.48	-45.71	2.95	-20.85	13.4	19.2	50.8	15.5	21
1985	8	15	.0	2446292.5	6 .424	+19 6.11	6 2.496	+19 6.16	3.45	-46.05	2.94	-20.88	13.4	19.1	51.7	15.7	34
1985	8	16	.0	2446293.5	6 .971	+19 6.99	6 3.045	+19 7.03	3.43	-46.39	2.92	-20.92	13.3	19.1	52.5	15.9	47
1985	8	17	.0	2446294.5	6 1.512	+19 7.88	6 3.586	+19 7.88	3.40	-46.72	2.91	-20.95	13.3	19.1	53.3	16.2	60
1985	8	18	.0	2446295.5	6 2.045	+19 8.76	6 4.120	+19 8.74	3.37	-47.04	2.90	-20.99	13.2	19.1	54.1	16.4	74
1985	8	19	.0	2446296.5	6 2.571	+19 9.64	6 4.646	+19 9.59	3.35	-47.36	2.89	-21.03	13.2	19.0	55.0	16.7	88
1985	8	20	.0	2446297.5	6 3.088	+19 10.51	6 5.164	+19 10.44	3.32	-47.68	2.87	-21.07	13.2	19.0	55.8	16.9	102
1985	8	21	.0	2446298.5	6 3.598	+19 11.38	6 5.674	+19 11.28	3.29	-47.99	2.86	-21.10	13.1	19.0	56.7	17.2	117
1985	8	22	.0	2446299.5	6 4.099	+19 12.26	6 6.176	+19 12.13	3.26	-48.30	2.85	-21.14	13.1	18.9	57.5	17.4	131
1985	8	23	.0	2446300.5	6 4.591	+19 13.13	6 6.668	+19 12.98	3.24	-48.60	2.84	-21.18	13.0	18.9	58.3	17.7	144
1985	8	24	.0	2446301.5	6 5.074	+19 14.00	6 7.152	+19 13.83	3.21	-48.89	2.83	-21.22	13.0	18.9	59.2	17.9	158
1985	8	25	.0	2446302.5	6 5.547	+19 14.88	6 7.627	+19 14.68	3.18	-49.18	2.81	-21.25	13.0	18.9	60.0	18.1	169
1985	8	26	.0	2446303.5	6 6.011	+19 15.76	6 8.091	+19 15.53	3.15	-49.47	2.80	-21.29	12.9	18.8	60.9	18.4	169
1985	8	27	.0	2446304.5	6 6.465	+19 16.64	6 8.545	+19 16.39	3.12	-49.75	2.79	-21.33	12.9	18.8	61.7	18.6	158
1985	8	28	.0	2446305.5	6 6.908	+19 17.53	6 8.989	+19 17.26	3.09	-50.03	2.78	-21.37	12.8	18.8	62.6	18.8	145
1985	8	29	.0	2446306.5	6 7.340	+19 18.42	6 9.422	+19 18.13	3.06	-50.30	2.76	-21.41	12.8	18.7	63.5	19.1	132
1985	8	30	.0	2446307.5	6 7.760	+19 19.32	6 9.843	+19 19.01	3.03	-50.57	2.75	-21.45	12.8	18.7	64.3	19.3	120
1985	8	31	.0	2446308.5	6 8.169	+19 20.23	6 10.253	+19 19.90	3.01	-50.83	2.74	-21.49	12.7	18.7	65.2	19.5	107
1985	9	1	.0	2446309.5	6 8.566	+19 21.15	6 10.650	+19 20.80	2.98	-51.09	2.73	-21.53	12.7	18.6	66.1	19.8	95
1985	9	2	.0	2446310.5	6 8.951	+19 22.07	6 11.035	+19 21.70	2.95	-51.34	2.71	-21.57	12.6	18.6	66.9	20.0	83
1985	9	3	.0	2446311.5	6 9.322	+19 23.01	6 11.407	+19 22.63	2.92	-51.59	2.70	-21.61	12.6	18.6	67.8	20.2	71
1985	9	4	.0	2446312.5	6 9.679	+19 23.97	6 11.765	+19 23.56	2.89	-51.83	2.69	-21.65	12.5	18.6	68.7	20.4	60
1985	9	5	.0	2446313.5	6 10.023	+19 24.93	6 12.109	+19 24.51	2.86	-52.07	2.68	-21.69	12.5	18.5	69.6	20.7	48
1985	9	6	.0	2446314.5	6 10.351	+19 25.92	6 12.438	+19 25.47	2.83	-52.30	2.66	-21.73	12.5	18.5	70.5	20.9	36
1985	9	7	.0	2446315.5	6 10.665	+19 26.92	6 12.753	+19 26.46	2.80	-52.52	2.65	-21.77	12.4	18.4	71.4	21.1	25
1985	9	8	.0	2446316.5	6 10.962	+19 27.93	6 13.051	+19 27.46	2.77	-52.74	2.64	-21.81	12.4	18.4	72.3	21.3	14
1985	9	9	.0	2446317.5	6 11.243	+19 28.97	6 13.332	+19 28.48	2.74	-52.96	2.63	-21.85	12.3	18.4	73.2	21.5	8
1985	9	10	.0	2446318.5	6 11.506	+19 30.03	6 13.597	+19 29.52	2.70	-53.17	2.61	-21.89	12.3	18.3	74.1	21.7	14
1985	9	11	.0	2446319.5	6 11.752	+19 31.11	6 13.843	+19 30.59	2.67	-53.37	2.60	-21.93	12.2	18.3	75.0	21.9	26

YR	MN	DY	HR	J.D.	R.A. (1950) .0	DEC.	R.A. APPN	DEC.	DELTA	DELDOT	R	RDOT	TMAG	NMAG	THETA	BETA	MOON
1985	9	12	.0	2446320.5	6 11.979	+19 32.22	6 14.071	+19 31.69	2.64	-53.57	2.59	-21.98	12.2	18.3	75.9	22.1	39
1985	9	13	.0	2446321.5	6 12.186	+19 33.35	6 14.279	+19 32.81	2.61	-53.75	2.58	-22.02	12.1	18.2	76.8	22.4	52
1985	9	14	.0	2446322.5	6 12.373	+19 34.51	6 14.467	+19 33.96	2.58	-53.94	2.56	-22.06	12.1	18.2	77.7	22.5	66
1985	9	15	.0	2446323.5	6 12.539	+19 35.70	6 14.633	+19 35.14	2.55	-54.11	2.55	-22.10	12.0	18.2	78.7	22.7	81
1985	9	16	.0	2446324.5	6 12.683	+19 36.92	6 14.778	+19 36.36	2.52	-54.28	2.54	-22.15	12.0	18.1	79.6	22.9	95
1985	9	17	.0	2446325.5	6 12.804	+19 38.17	6 14.899	+19 37.61	2.49	-54.44	2.53	-22.19	11.9	18.1	80.6	23.1	110
1985	9	18	.0	2446326.5	6 12.901	+19 39.47	6 14.997	+19 38.89	2.46	-54.59	2.51	-22.24	11.9	18.1	81.5	23.3	124
1985	9	19	.0	2446327.5	6 12.973	+19 40.80	6 15.070	+19 40.22	2.42	-54.73	2.50	-22.28	11.8	18.0	82.5	23.5	139
1985	9	20	.0	2446328.5	6 13.019	+19 42.17	6 15.118	+19 41.59	2.39	-54.86	2.49	-22.32	11.8	18.0	83.4	23.7	153
1985	9	21	.0	2446329.5	6 13.039	+19 43.58	6 15.138	+19 43.00	2.36	-54.99	2.47	-22.37	11.7	17.9	84.4	23.8	165
1985	9	22	.0	2446330.5	6 13.030	+19 45.04	6 15.131	+19 44.46	2.33	-55.11	2.46	-22.41	11.6	17.9	85.4	24.0	171
1985	9	23	.0	2446331.5	6 12.992	+19 46.55	6 15.094	+19 45.97	2.30	-55.22	2.45	-22.46	11.6	17.8	86.4	24.1	162
1985	9	24	.0	2446332.5	6 12.924	+19 48.11	6 15.027	+19 47.53	2.27	-55.32	2.43	-22.50	11.5	17.8	87.4	24.3	150
1985	9	25	.0	2446333.5	6 12.825	+19 49.72	6 14.928	+19 49.15	2.23	-55.41	2.42	-22.55	11.5	17.8	88.4	24.5	137
1985	9	26	.0	2446334.5	6 12.692	+19 51.39	6 14.797	+19 50.82	2.20	-55.50	2.41	-22.59	11.4	17.7	89.4	24.6	124
1985	9	27	.0	2446335.5	6 12.525	+19 53.11	6 14.630	+19 52.56	2.17	-55.58	2.40	-22.64	11.4	17.7	90.4	24.7	112
1985	9	28	.0	2446336.5	6 12.321	+19 54.90	6 14.428	+19 54.35	2.14	-55.65	2.38	-22.69	11.3	17.6	91.4	24.9	100
1985	9	29	.0	2446337.5	6 12.080	+19 56.75	6 14.188	+19 56.21	2.10	-55.71	2.37	-22.73	11.2	17.6	92.5	25.0	87
1985	9	30	.0	2446338.5	6 11.800	+19 58.66	6 13.908	+19 58.15	2.07	-55.76	2.36	-22.78	11.2	17.5	93.5	25.1	75
1985	10	1	.0	2446339.5	6 11.478	+20 .65	6 13.587	+20 .15	2.04	-55.80	2.34	-22.83	11.1	17.5	94.6	25.2	63
1985	10	2	.0	2446340.5	6 11.113	+20 2.71	6 13.223	+20 2.22	2.01	-55.83	2.33	-22.88	11.1	17.5	95.6	25.3	52
1985	10	3	.0	2446341.5	6 10.702	+20 4.84	6 12.814	+20 4.38	1.98	-55.85	2.32	-22.92	11.0	17.4	96.7	25.4	40
1985	10	4	.0	2446342.5	6 10.245	+20 7.06	6 12.357	+20 6.61	1.94	-55.86	2.30	-22.97	10.9	17.4	97.8	25.5	28
1985	10	5	.0	2446343.5	6 9.737	+20 9.35	6 11.851	+20 8.94	1.91	-55.86	2.29	-23.02	10.9	17.3	98.9	25.6	17
1985	10	6	.0	2446344.5	6 9.176	+20 11.73	6 11.292	+20 11.34	1.88	-55.85	2.28	-23.07	10.8	17.3	100.0	25.6	8
1985	10	7	.0	2446345.5	6 8.561	+20 14.20	6 10.678	+20 13.84	1.85	-55.83	2.26	-23.12	10.7	17.2	101.1	25.7	11
1985	10	8	.0	2446346.5	6 7.887	+20 16.76	6 10.006	+20 16.44	1.81	-55.79	2.25	-23.17	10.7	17.2	102.3	25.7	22
1985	10	9	.0	2446347.5	6 7.153	+20 19.42	6 9.272	+20 19.13	1.78	-55.74	2.24	-23.22	10.6	17.1	103.5	25.7	35
1985	10	10	.0	2446348.5	6 6.354	+20 22.17	6 8.475	+20 21.92	1.75	-55.68	2.22	-23.26	10.5	17.1	104.6	25.8	48
1985	10	11	.0	2446349.5	6 5.488	+20 25.02	6 7.610	+20 24.82	1.72	-55.60	2.21	-23.31	10.5	17.0	105.8	25.8	62
1985	10	12	.0	2446350.5	6 4.550	+20 27.97	6 6.673	+20 27.83	1.69	-55.50	2.20	-23.36	10.4	16.9	107.0	25.8	76
1985	10	13	.0	2446351.5	6 3.536	+20 31.04	6 5.661	+20 30.94	1.65	-55.39	2.18	-23.42	10.3	16.9	108.3	25.7	91
1985	10	14	.0	2446352.5	6 2.444	+20 34.21	6 4.570	+20 34.17	1.62	-55.26	2.17	-23.47	10.3	16.8	109.5	25.7	106
1985	10	15	.0	2446353.5	6 1.267	+20 37.49	6 3.395	+20 37.51	1.59	-55.12	2.16	-23.52	10.2	16.8	110.8	25.6	121
1985	10	16	.0	2446354.5	6 .002	+20 40.88	6 2.131	+20 40.97	1.56	-54.95	2.14	-23.57	10.1	16.7	112.1	25.5	136
1985	10	17	.0	2446355.5	5 58.643	+20 44.39	6 .774	+20 44.55	1.53	-54.77	2.13	-23.62	10.0	16.7	113.4	25.5	151
1985	10	18	.0	2446356.5	5 57.185	+20 48.02	5 59.318	+20 48.24	1.50	-54.56	2.11	-23.67	10.0	16.6	114.7	25.3	166
1985	10	19	.0	2446357.5	5 55.623	+20 51.76	5 57.757	+20 52.06	1.46	-54.33	2.10	-23.72	9.9	16.5	116.1	25.2	172
1985	10	20	.0	2446358.5	5 53.949	+20 55.61	5 56.086	+20 56.00	1.43	-54.08	2.09	-23.78	9.8	16.5	117.4	25.0	162
1985	10	21	.0	2446359.5	5 52.159	+20 59.57	5 54.297	+21 .06	1.40	-53.81	2.07	-23.83	9.7	16.4	118.9	24.9	148
1985	10	22	.0	2446360.5	5 50.244	+21 3.65	5 52.383	+21 4.24	1.37	-53.51	2.06	-23.88	9.6	16.4	120.3	24.7	135
1985	10	23	.0	2446361.5	5 48.197	+21 7.83	5 50.338	+21 8.52	1.34	-53.19	2.05	-23.94	9.6	16.3	121.8	24.4	122
1985	10	24	.0	2446362.5	5 46.009	+21 12.11	5 48.152	+21 12.91	1.31	-52.84	2.03	-23.99	9.5	16.2	123.3	24.2	109
1985	10	25	.0	2446363.5	5 43.673	+21 16.47	5 45.818	+21 17.40	1.28	-52.45	2.02	-24.04	9.4	16.2	124.8	23.9	96
1985	10	26	.0	2446364.5	5 41.179	+21 20.92	5 43.325	+21 21.98	1.25	-52.04	2.00	-24.10	9.3	16.1	126.4	23.5	83
1985	10	27	.0	2446365.5	5 38.517	+21 25.43	5 40.664	+21 26.63	1.22	-51.59	1.99	-24.15	9.2	16.0	128.0	23.2	71
1985	10	28	.0	2446366.5	5 35.675	+21 29.99	5 37.824	+21 31.33	1.19	-51.10	1.98	-24.20	9.1	16.0	129.7	22.8	58
1985	10	29	.0	2446367.5	5 32.644	+21 34.57	5 34.794	+21 36.07	1.16	-50.58	1.96	-24.26	9.0	15.9	131.4	22.3	46
1985	10	30	.0	2446368.5	5 29.409	+21 39.16	5 31.561	+21 40.82	1.13	-50.01	1.95	-24.31	9.0	15.8	133.1	21.8	33
1985	10	31	.0	2446369.5	5 25.960	+21 43.71	5 28.113	+21 45.55	1.10	-49.40	1.93	-24.37	8.9	15.7	134.9	21.3	21

YR	MN	DY	HR	J.D.	R.A. (1950) .0	DEC.	R.A. APPN	DEC.	DELTA	DELDOT	R	RDOT	TMAG	NMAG	THETA	BETA	MOON
1985	11	1	.0	2446370.5	5 22.281	+21 48.18	5 24.435	+21 50.22	1.07	-48.73	1.92	-24.42	8.8	15.7	136.8	20.7	9
1985	11	2	.0	2446371.5	5 18.358	+21 52.55	5 20.513	+21 54.78	1.05	-48.02	1.91	-24.48	8.7	15.6	138.7	20.1	6
1985	11	3	.0	2446372.5	5 14.175	+21 56.73	5 16.331	+21 59.18	1.02	-47.24	1.89	-24.54	8.6	15.5	140.6	19.4	18
1985	11	4	.0	2446373.5	5 9.717	+22 .69	5 11.874	+22 3.36	.99	-46.40	1.88	-24.59	8.5	15.4	142.7	18.7	31
1985	11	5	.0	2446374.5	5 4.968	+22 4.33	5 7.124	+22 7.25	.96	-45.50	1.86	-24.65	8.4	15.4	144.7	17.9	44
1985	11	6	.0	2446375.5	4 55.909	+22 7.57	5 2.065	+22 10.74	.94	-44.52	1.85	-24.70	8.3	15.3	146.9	17.0	58
1985	11	7	.0	2446376.5	4 54.523	+22 10.30	4 56.679	+22 13.75	.91	-43.46	1.84	-24.76	8.2	15.2	149.2	16.1	72
1985	11	8	.0	2446377.5	4 48.793	+22 12.41	4 50.948	+22 16.15	.89	-42.32	1.82	-24.82	8.1	15.1	151.5	15.1	87
1985	11	9	.0	2446378.5	4 42.702	+22 13.77	4 44.854	+22 17.80	.86	-41.09	1.81	-24.87	8.0	15.1	153.9	14.0	103
1985	11	10	.0	2446379.5	4 36.232	+22 14.20	4 38.382	+22 19.56	.84	-39.76	1.79	-24.93	7.9	15.0	156.4	12.8	119
1985	11	11	.0	2446380.5	4 29.369	+22 13.56	4 31.516	+22 18.25	.82	-38.33	1.78	-24.99	7.8	14.9	158.9	11.5	135
1985	11	12	.0	2446381.5	4 22.098	+22 11.63	4 24.241	+22 16.67	.80	-36.79	1.76	-25.04	7.7	14.8	161.6	10.2	152
1985	11	13	.0	2446382.5	4 14.410	+22 8.21	4 16.547	+22 13.62	.78	-35.13	1.75	-25.10	7.6	14.8	164.4	8.8	169
1985	11	14	.0	2446383.5	4 6.295	+22 3.08	4 8.426	+22 8.86	.76	-33.36	1.73	-25.16	7.5	14.7	167.2	7.3	173
1985	11	15	.0	2446384.5	3 57.750	+21 55.99	3 59.875	+22 2.16	.74	-31.47	1.72	-25.21	7.4	14.6	170.1	5.7	156
1985	11	16	.0	2446385.5	3 48.778	+21 46.68	3 50.895	+21 53.26	.72	-29.45	1.71	-25.27	7.3	14.5	173.1	4.0	139
1985	11	17	.0	2446386.5	3 39.387	+21 34.93	3 41.493	+21 41.91	.70	-27.30	1.69	-25.33	7.2	14.5	176.0	2.4	123
1985	11	18	.0	2446387.5	3 29.590	+21 20.46	3 31.686	+21 27.86	.69	-25.03	1.68	-25.38	7.1	14.4	177.7	1.4	107
1985	11	19	.0	2446388.5	3 19.411	+21 3.08	3 21.495	+21 10.89	.67	-22.64	1.66	-25.44	7.1	14.3	175.9	2.4	92
1985	11	20	.0	2446389.5	3 8.881	+20 42.57	3 10.952	+20 50.80	.66	-20.13	1.65	-25.50	7.0	14.3	172.8	4.3	76
1985	11	21	.0	2446390.5	2 58.040	+20 18.80	3 .096	+20 27.44	.65	-17.52	1.63	-25.55	6.9	14.2	169.3	6.4	61
1985	11	22	.0	2446391.5	2 46.935	+19 51.67	2 48.975	+20 .71	.64	-14.81	1.62	-25.61	6.8	14.2	165.7	8.7	47
1985	11	23	.0	2446392.5	2 35.621	+19 21.16	2 37.645	+19 30.59	.63	-12.03	1.60	-25.67	6.8	14.1	162.0	11.0	32
1985	11	24	.0	2446393.5	2 24.161	+18 47.34	2 26.168	+18 57.13	.63	-9.19	1.59	-25.72	6.7	14.1	158.3	13.3	18
1985	11	25	.0	2446394.5	2 12.619	+18 10.35	2 14.609	+18 20.48	.62	-6.31	1.57	-25.78	6.6	14.1	154.5	15.7	5
1985	11	26	.0	2446395.5	2 1.065	+17 30.42	2 3.037	+17 40.86	.62	-3.43	1.56	-25.83	6.6	14.0	150.7	18.1	12
1985	11	27	.0	2446396.5	1 49.566	+16 47.86	1 51.522	+16 58.58	.62	-.55	1.54	-25.89	6.5	14.0	146.9	20.5	26
1985	11	28	.0	2446397.5	1 38.189	+16 3.02	1 40.129	+16 14.01	.62	2.29	1.53	-25.94	6.5	14.0	143.0	22.8	41
1985	11	29	.0	2446398.5	1 26.996	+15 16.35	1 28.920	+15 27.56	.62	5.07	1.51	-25.99	6.4	14.0	139.2	25.2	55
1985	11	30	.0	2446399.5	1 16.043	+14 28.29	1 17.953	+14 39.69	.63	7.77	1.50	-26.05	6.4	14.0	135.5	27.5	70
1985	12	1	.0	2446400.5	1 5.379	+13 39.29	1 7.275	+13 50.87	.63	10.37	1.48	-26.10	6.4	14.0	131.7	29.7	85
1985	12	2	.0	2446401.5	0 55.043	+12 49.83	0 56.926	+13 1.54	.64	12.85	1.47	-26.15	6.3	14.0	128.1	31.9	100
1985	12	3	.0	2446402.5	0 45.067	+12 .33	0 46.940	+12 12.14	.65	15.20	1.45	-26.20	6.3	14.0	124.5	34.0	115
1985	12	4	.0	2446403.5	0 35.476	+11 11.18	0 37.339	+11 23.07	.66	17.40	1.44	-26.25	6.3	14.0	121.0	36.0	130
1985	12	5	.0	2446404.5	0 26.284	+10 22.72	0 28.138	+10 34.68	.67	19.47	1.42	-26.30	6.3	14.0	117.6	37.9	145
1985	12	6	.0	2446405.5	0 17.500	+ 9 35.26	0 19.347	+ 9 47.25	.68	21.38	1.41	-26.34	6.3	14.0	114.3	39.7	159
1985	12	7	.0	2446406.5	0 9.127	+ 8 49.03	0 10.967	+ 9 1.04	.69	23.14	1.39	-26.39	6.3	14.0	111.1	41.3	169
1985	12	8	.0	2446407.5	0 1.161	+ 8 4.22	0 2.996	+ 8 16.23	.70	24.75	1.38	-26.43	6.2	14.0	108.0	42.9	162
1985	12	9	.0	2446408.5	23 53.595	+ 7 20.97	23 55.425	+ 7 32.97	.72	26.22	1.36	-26.48	6.2	14.1	105.0	44.3	147
1985	12	10	.0	2446409.5	23 46.417	+ 6 39.38	23 48.244	+ 6 51.36	.73	27.55	1.35	-26.52	6.2	14.1	102.1	45.7	131
1985	12	11	.0	2446410.5	23 39.614	+ 5 59.50	23 41.439	+ 6 11.45	.75	28.75	1.33	-26.56	6.2	14.1	99.2	46.9	114
1985	12	12	.0	2446411.5	23 33.172	+ 5 21.37	23 34.994	+ 5 33.27	.77	29.81	1.32	-26.59	6.2	14.1	96.5	48.1	97
1985	12	13	.0	2446412.5	23 27.072	+ 4 44.97	23 28.893	+ 4 56.83	.79	30.76	1.30	-26.63	6.2	14.1	93.9	49.1	81
1985	12	14	.0	2446413.5	23 21.298	+ 4 10.29	23 23.118	+ 4 22.09	.80	31.59	1.28	-26.66	6.2	14.2	91.3	50.0	65
1985	12	15	.0	2446414.5	23 15.832	+ 3 37.27	23 17.652	+ 3 49.02	.82	32.31	1.27	-26.69	6.2	14.2	88.8	50.8	50
1985	12	16	.0	2446415.5	23 10.656	+ 3 5.88	23 12.475	+ 3 17.57	.84	32.94	1.25	-26.72	6.2	14.2	86.4	51.6	35
1985	12	17	.0	2446416.5	23 5.752	+ 2 36.04	23 7.572	+ 2 47.67	.86	33.47	1.24	-26.74	6.2	14.2	84.1	52.2	22
1985	12	18	.0	2446417.5	23 1.104	+ 2 7.69	23 2.924	+ 2 19.25	.88	33.92	1.22	-26.76	6.2	14.3	81.8	52.8	12
1985	12	19	.0	2446418.5	22 56.695	+ 1 40.75	22 58.515	+ 1 52.25	.90	34.29	1.21	-26.78	6.1	14.3	79.6	53.3	14
1985	12	20	.0	2446419.5	22 52.509	+ 1 15.15	22 54.330	+ 1 26.58	.92	34.58	1.19	-26.79	6.1	14.3	77.5	53.7	24

YR	MN	DY	HR	J.D.	R.A. (1950)		.0 DEC.			R.A.		APPN DEC.			DELTA	DELDOT	R	RDOT	TMAG	NMAG	THETA	BETA	MOON
1985	12	21	.0	2446420.5	22	48.532	+	0	50.82	22	50.353	+	1	2.18	.94	34.80	1.18	-26.80	6.1	14.3	75.4	54.0	36
1985	12	22	.0	2446421.5	22	44.748	+	0	27.67	22	46.571	+	0	38.96	.96	34.96	1.16	-26.81	6.1	14.3	73.4	54.3	48
1985	12	23	.0	2446422.5	22	41.145	+	0	5.63	22	42.969	+	0	16.86	.98	35.06	1.15	-26.81	6.1	14.3	71.4	54.5	61
1985	12	24	.0	2446423.5	22	37.710	-	0	15.36	22	39.535	-	0	4.20	1.00	35.10	1.13	-26.80	6.1	14.4	69.5	54.6	73
1985	12	25	.0	2446424.5	22	34.431	-	0	35.37	22	36.257	-	0	24.28	1.02	35.09	1.11	-26.79	6.0	14.4	67.6	54.7	86
1985	12	26	.0	2446425.5	22	31.296	-	0	54.48	22	33.124	-	0	43.45	1.04	35.02	1.10	-26.77	6.0	14.4	65.7	54.7	98
1985	12	27	.0	2446426.5	22	28.296	-	1	12.74	22	30.125	-	1	1.77	1.06	34.91	1.08	-26.74	6.0	14.4	63.9	54.6	111
1985	12	28	.0	2446427.5	22	25.420	-	1	30.22	22	27.251	-	1	19.31	1.08	34.76	1.07	-26.71	6.0	14.4	62.1	54.5	124
1985	12	29	.0	2446428.5	22	22.660	-	1	46.96	22	24.492	-	1	36.12	1.10	34.56	1.05	-26.67	5.9	14.4	60.4	54.3	136
1985	12	30	.0	2446429.5	22	20.006	-	2	3.03	22	21.839	-	1	52.25	1.12	34.31	1.04	-26.62	5.9	14.4	58.7	54.1	149
1985	12	31	.0	2446430.5	22	17.450	-	2	18.48	22	19.285	-	2	7.75	1.14	34.03	1.02	-26.56	5.9	14.4	57.0	53.8	160
1986	1	1	.0	2446431.5	22	14.985	-	2	33.34	22	16.822	-	2	22.68	1.16	33.71	1.01	-26.49	5.8	14.4	55.3	53.4	167
1986	1	2	.0	2446432.5	22	12.603	-	2	47.69	22	14.442	-	2	37.08	1.18	33.35	.99	-26.41	5.8	14.4	53.7	53.0	162
1986	1	3	.0	2446433.5	22	10.299	-	3	1.54	22	12.138	-	2	50.99	1.20	32.96	.98	-26.32	5.7	14.4	52.0	52.6	150
1986	1	4	.0	2446434.5	22	8.064	-	3	14.95	22	9.905	-	3	4.46	1.22	32.52	.96	-26.22	5.7	14.4	50.4	52.1	137
1986	1	5	.0	2446435.5	22	5.894	-	3	27.96	22	7.737	-	3	17.52	1.24	32.05	.95	-26.10	5.7	14.4	48.9	51.5	123
1986	1	6	.0	2446436.5	22	3.783	-	3	40.60	22	5.627	-	3	30.22	1.25	31.55	.93	-25.96	5.6	14.4	47.3	50.9	108
1986	1	7	.0	2446437.5	22	1.726	-	3	52.90	22	3.571	-	3	42.58	1.27	31.01	.92	-25.81	5.6	14.4	45.7	50.3	94
1986	1	8	.0	2446438.5	21	59.716	-	4	4.92	22	1.564	-	3	54.65	1.29	30.43	.90	-25.64	5.5	14.4	44.2	49.5	79
1986	1	9	.0	2446439.5	21	57.750	-	4	16.66	21	59.599	-	4	6.45	1.31	29.81	.89	-25.46	5.5	14.4	42.7	48.8	64
1986	1	10	.0	2446440.5	21	55.823	-	4	28.18	21	57.674	-	4	18.02	1.32	29.16	.87	-25.25	5.4	14.4	41.2	48.0	49
1986	1	11	.0	2446441.5	21	53.930	-	4	39.50	21	55.783	-	4	29.39	1.34	28.47	.86	-25.02	5.4	14.4	39.7	47.1	35
1986	1	12	.0	2446442.5	21	52.068	-	4	50.64	21	53.923	-	4	40.58	1.36	27.75	.84	-24.76	5.3	14.4	38.2	46.2	22
1986	1	13	.0	2446443.5	21	50.232	-	5	1.64	21	52.088	-	4	51.64	1.37	26.99	.83	-24.48	5.3	14.4	36.7	45.2	12
1986	1	14	.0	2446444.5	21	48.419	-	5	12.53	21	50.277	-	5	2.58	1.39	26.19	.81	-24.17	5.2	14.4	35.2	44.1	15
1986	1	15	.0	2446445.5	21	46.624	-	5	23.34	21	48.484	-	5	13.44	1.40	25.35	.80	-23.83	5.1	14.4	33.7	43.0	25
1986	1	16	.0	2446446.5	21	44.846	-	5	34.08	21	46.707	-	5	24.23	1.42	24.48	.79	-23.46	5.1	14.3	32.3	41.9	37
1986	1	17	.0	2446447.5	21	43.080	-	5	44.78	21	44.943	-	5	34.99	1.43	23.56	.77	-23.06	5.0	14.3	30.8	40.7	50
1986	1	18	.0	2446448.5	21	41.323	-	5	55.47	21	43.188	-	5	45.74	1.44	22.60	.76	-22.62	4.9	14.3	29.4	39.4	62
1986	1	19	.0	2446449.5	21	39.574	-	6	6.18	21	41.440	-	5	56.50	1.46	21.60	.75	-22.14	4.9	14.3	27.9	38.1	74
1986	1	20	.0	2446450.5	21	37.828	-	6	16.92	21	39.697	-	6	7.29	1.47	20.56	.73	-21.62	4.8	14.2	26.5	36.7	86
1986	1	21	.0	2446451.5	21	36.085	-	6	27.72	21	37.956	-	6	18.14	1.48	19.47	.72	-21.05	4.8	14.2	25.1	35.3	98
1986	1	22	.0	2446452.5	21	34.342	-	6	38.59	21	36.214	-	6	29.07	1.49	18.34	.71	-20.44	4.7	14.2	23.6	33.8	110
1986	1	23	.0	2446453.5	21	32.596	-	6	49.57	21	34.471	-	6	40.10	1.50	17.17	.70	-19.78	4.6	14.2	22.2	32.2	122
1986	1	24	.0	2446454.5	21	30.847	-	7	.66	21	32.724	-	6	51.25	1.51	15.94	.69	-19.07	4.6	14.2	20.8	30.6	134
1986	1	25	.0	2446455.5	21	29.092	-	7	11.89	21	30.972	-	7	2.54	1.52	14.67	.68	-18.31	4.5	14.2	19.4	28.9	147
1986	1	26	.0	2446456.5	21	27.331	-	7	23.27	21	29.213	-	7	13.98	1.53	13.35	.67	-17.49	4.4	14.1	18.0	27.2	158
1986	1	27	.0	2446457.5	21	25.562	-	7	34.83	21	27.446	-	7	25.59	1.54	11.99	.66	-16.61	4.4	14.1	16.7	25.5	167
1986	1	28	.0	2446458.5	21	23.784	-	7	46.57	21	25.671	-	7	37.40	1.54	10.57	.65	-15.68	4.3	14.1	15.3	23.7	164
1986	1	29	.0	2446459.5	21	21.998	-	7	58.51	21	23.887	-	7	49.40	1.55	9.11	.64	-14.70	4.3	14.1	14.0	21.9	153
1986	1	30	.0	2446460.5	21	20.203	-	8	10.67	21	22.094	-	8	1.62	1.55	7.61	.63	-13.65	4.2	14.1	12.7	20.0	140
1986	1	31	.0	2446461.5	21	18.398	-	8	23.05	21	20.292	-	8	14.06	1.56	6.06	.62	-12.55	4.1	14.0	11.4	18.2	127
1986	2	1	.0	2446462.5	21	16.585	-	8	35.66	21	18.482	-	8	26.74	1.56	4.47	.62	-11.39	4.1	14.0	10.2	16.5	113
1986	2	2	.0	2446463.5	21	14.763	-	8	48.52	21	16.663	-	8	39.66	1.56	2.84	.61	-10.18	4.1	14.0	9.1	14.8	99
1986	2	3	.0	2446464.5	21	12.934	-	9	1.64	21	14.837	-	8	52.83	1.56	1.18	.60	-8.93	4.0	14.0	8.1	13.3	84
1986	2	4	.0	2446465.5	21	11.099	-	9	15.00	21	13.004	-	9	6.27	1.56	-.52	.60	-7.62	4.0	14.0	7.3	12.1	70
1986	2	5	.0	2446466.5	21	9.258	-	9	28.63	21	11.167	-	9	19.96	1.56	-2.24	.60	-6.28	3.9	13.9	6.8	11.2	55
1986	2	6	.0	2446467.5	21	7.413	-	9	42.52	21	9.326	-	9	33.92	1.56	-3.98	.59	-4.91	3.9	13.9	6.5	10.9	41
1986	2	7	.0	2446468.5	21	5.566	-	9	56.68	21	7.482	-	9	48.14	1.56	-5.73	.59	-3.50	3.9	13.9	6.6	11.1	27
1986	2	8	.0	2446469.5	21	3.718	-10		11.11	21	5.638	-10		2.63	1.55	-7.48	.59	-2.08	3.9	13.9	7.0	11.8	15

YR	MN	DY	HR	J.D.	R.A. (1950) .0 h	m	DEC. °	'	R.A. APPN h	m	DEC. °	'	DELTA	DELDOT	R	RDOT	TMAG	NMAG	THETA	BETA	MOON
1986	2	9	.0	2446470.5	21	1.872	-10	25.80	21	3.794	-10	17.39	1.55	-9.24	.59	-.65	4.1	13.9	7.7	13.1	11
1986	2	10	.0	2446471.5	21	.028	-10	40.75	21	1.954	-10	32.42	1.54	-10.99	.59	.79	4.1	13.9	8.6	14.6	20
1986	2	11	.0	2446472.5	20	58.188	-10	55.97	21	.118	-10	47.71	1.54	-12.72	.59	2.22	4.1	13.9	9.7	16.4	32
1986	2	12	.0	2446473.5	20	56.354	-11	11.46	20	58.288	-11	3.26	1.53	-14.43	.59	3.64	4.1	13.9	10.9	18.4	45
1986	2	13	.0	2446474.5	20	54.528	-11	27.20	20	56.466	-11	19.07	1.52	-16.12	.59	5.04	4.1	13.9	12.1	20.5	58
1986	2	14	.0	2446475.5	20	52.711	-11	43.21	20	54.652	-11	35.15	1.51	-17.77	.60	6.41	4.1	13.9	13.4	22.6	70
1986	2	15	.0	2446476.5	20	50.904	-11	59.47	20	52.849	-11	51.48	1.50	-19.38	.60	7.75	4.1	13.9	14.7	24.8	83
1986	2	16	.0	2446477.5	20	49.107	-12	16.00	20	51.056	-12	8.08	1.49	-20.95	.60	9.05	4.1	13.9	16.1	26.9	95
1986	2	17	.0	2446478.5	20	47.322	-12	32.80	20	49.275	-12	24.94	1.48	-22.46	.61	10.30	4.1	13.9	17.5	29.1	107
1986	2	18	.0	2446479.5	20	45.548	-12	49.86	20	47.506	-12	42.08	1.46	-23.93	.62	11.51	4.2	13.9	18.9	31.2	119
1986	2	19	.0	2446480.5	20	43.786	-13	7.21	20	45.748	-12	59.49	1.45	-25.35	.62	12.66	4.2	13.9	20.3	33.3	132
1986	2	20	.0	2446481.5	20	42.035	-13	24.83	20	44.002	-13	17.18	1.43	-26.70	.63	13.75	4.2	13.9	21.7	35.4	144
1986	2	21	.0	2446482.5	20	40.294	-13	42.76	20	42.266	-13	35.17	1.42	-28.01	.64	14.79	4.2	13.9	23.1	37.4	156
1986	2	22	.0	2446483.5	20	38.562	-14	.99	20	40.539	-13	53.48	1.40	-29.25	.65	15.78	4.2	13.9	24.6	39.4	167
1986	2	23	.0	2446484.5	20	36.838	-14	19.55	20	38.819	-14	12.10	1.38	-30.44	.66	16.70	4.2	13.9	26.0	41.3	169
1986	2	24	.0	2446485.5	20	35.119	-14	38.46	20	37.106	-14	31.08	1.37	-31.57	.67	17.57	4.3	13.9	27.4	43.1	159
1986	2	25	.0	2446486.5	20	33.404	-14	57.73	20	35.396	-14	50.42	1.35	-32.65	.68	18.38	4.3	13.9	28.9	44.9	146
1986	2	26	.0	2446487.5	20	31.690	-15	17.40	20	33.687	-15	10.16	1.33	-33.67	.69	19.14	4.3	13.9	30.3	46.6	132
1986	2	27	.0	2446488.5	20	29.974	-15	37.49	20	31.976	-15	30.32	1.31	-34.64	.70	19.85	4.3	13.9	31.8	48.2	118
1986	2	28	.0	2446489.5	20	28.252	-15	58.03	20	30.259	-15	50.93	1.29	-35.56	.71	20.50	4.4	13.9	33.3	49.8	103
1986	3	1	.0	2446490.5	20	26.520	-16	19.06	20	28.534	-16	12.03	1.27	-36.42	.72	21.11	4.4	13.9	34.7	51.3	88
1986	3	2	.0	2446491.5	20	24.775	-16	40.63	20	26.794	-16	33.67	1.25	-37.24	.74	21.67	4.4	13.9	36.2	52.7	74
1986	3	3	.0	2446492.5	20	23.012	-17	2.76	20	25.037	-16	55.87	1.22	-38.01	.75	22.19	4.4	13.9	37.7	54.1	59
1986	3	4	.0	2446493.5	20	21.225	-17	25.52	20	23.257	-17	18.70	1.20	-38.73	.76	22.66	4.4	13.9	39.2	55.3	45
1986	3	5	.0	2446494.5	20	19.408	-17	48.95	20	21.447	-17	42.21	1.18	-39.40	.77	23.10	4.4	13.9	40.7	56.6	31
1986	3	6	.0	2446495.5	20	17.556	-18	13.12	20	19.602	-18	6.45	1.16	-40.03	.79	23.50	4.5	13.9	42.2	57.7	17
1986	3	7	.0	2446496.5	20	15.662	-18	38.08	20	17.714	-18	31.49	1.13	-40.62	.80	23.87	4.5	13.9	43.7	58.8	6
1986	3	8	.0	2446497.5	20	13.717	-19	3.90	20	15.777	-18	57.40	1.11	-41.16	.82	24.20	4.5	13.9	45.3	59.8	13
1986	3	9	.0	2446498.5	20	11.713	-19	30.66	20	13.781	-19	24.25	1.09	-41.65	.83	24.51	4.5	13.9	46.8	60.7	27
1986	3	10	.0	2446499.5	20	9.642	-19	58.45	20	11.718	-19	52.12	1.06	-42.10	.84	24.79	4.5	13.9	48.4	61.6	40
1986	3	11	.0	2446500.5	20	7.493	-20	27.34	20	9.578	-20	21.10	1.04	-42.51	.86	25.04	4.5	13.8	50.0	62.4	54
1986	3	12	.0	2446501.5	20	5.255	-20	57.44	20	7.348	-20	51.30	1.01	-42.87	.87	25.27	4.5	13.8	51.6	63.1	67
1986	3	13	.0	2446502.5	20	2.916	-21	28.84	20	5.018	-21	22.80	.99	-43.18	.89	25.47	4.5	13.8	53.2	63.8	80
1986	3	14	.0	2446503.5	20	.461	-22	1.67	20	2.573	-21	55.74	.96	-43.44	.90	25.66	4.5	13.8	54.9	64.3	93
1986	3	15	.0	2446504.5	19	57.875	-22	36.05	19	59.997	-22	30.23	.94	-43.65	.92	25.83	4.5	13.8	56.6	64.8	105
1986	3	16	.0	2446505.5	19	55.140	-23	12.10	19	57.274	-23	6.40	.91	-43.81	.93	25.98	4.5	13.7	58.4	65.3	118
1986	3	17	.0	2446506.5	19	52.237	-23	49.97	19	54.384	-23	44.40	.89	-43.92	.95	26.11	4.5	13.7	60.1	65.6	131
1986	3	18	.0	2446507.5	19	49.144	-24	29.81	19	51.303	-24	24.38	.86	-43.97	.96	26.23	4.5	13.7	62.0	65.9	143
1986	3	19	.0	2446508.5	19	45.836	-25	11.79	19	48.009	-25	6.50	.84	-43.95	.98	26.33	4.5	13.7	63.8	66.1	156
1986	3	20	.0	2446509.5	19	42.282	-25	56.08	19	44.470	-25	50.96	.81	-43.87	.99	26.42	4.5	13.6	65.8	66.1	169
1986	3	21	.0	2446510.5	19	38.451	-26	42.86	19	40.655	-26	37.92	.79	-43.71	1.01	26.50	4.4	13.6	67.8	66.1	176
1986	3	22	.0	2446511.5	19	34.305	-27	32.33	19	36.526	-27	27.58	.76	-43.47	1.02	26.57	4.4	13.6	69.8	66.0	163
1986	3	23	.0	2446512.5	19	29.798	-28	24.68	19	32.038	-28	20.14	.73	-43.14	1.04	26.63	4.4	13.5	71.9	65.8	149
1986	3	24	.0	2446513.5	19	24.882	-29	20.11	19	27.142	-29	15.80	.71	-42.72	1.05	26.68	4.4	13.5	74.1	65.5	134
1986	3	25	.0	2446514.5	19	19.495	-30	18.79	19	21.777	-30	14.74	.69	-42.18	1.07	26.72	4.3	13.4	76.5	65.0	119
1986	3	26	.0	2446515.5	19	13.571	-31	20.90	19	15.876	-31	17.14	.66	-41.52	1.09	26.75	4.3	13.4	78.9	64.4	103
1986	3	27	.0	2446516.5	19	7.028	-32	26.57	19	9.359	-32	23.12	.64	-40.73	1.10	26.77	4.3	13.3	81.4	63.7	87
1986	3	28	.0	2446517.5	18	59.774	-33	35.86	19	2.133	-33	32.77	.61	-39.78	1.12	26.79	4.2	13.3	84.0	62.8	71
1986	3	29	.0	2446518.5	18	51.700	-34	48.75	18	54.089	-34	46.06	.59	-38.66	1.13	26.80	4.2	13.2	86.8	61.8	55
1986	3	30	.0	2446519.5	18	42.681	-36	5.09	18	45.102	-36	2.85	.57	-37.33	1.15	26.81	4.2	13.2	89.7	60.5	39

YR	MN	DY	HR	J.D.	R.A. (1950) .0 DEC.				R.A. APPN DEC.				DELTA	DELDOT	R	RDOT	TMAG	NMAG	THETA	BETA	MOON
1986	3	31	.0	2446520.5	18	32.573	-37	24.52	18	35.029	-37	22.80	.55	-35.79	1.16	26.81	4.1	13.1	92.7	59.1	23
1986	4	1	.0	2446521.5	18	21.216	-38	46.42	18	23.709	-38	45.29	.53	-34.01	1.18	26.80	4.1	13.1	96.0	57.5	11
1986	4	2	.0	2446522.5	18	8.434	-40	9.82	18	10.964	-40	9.35	.51	-31.95	1.19	26.79	4.1	13.0	99.4	55.7	16
1986	4	3	.0	2446523.5	17	54.041	-41	33.29	17	56.608	-41	33.57	.49	-29.59	1.21	26.78	4.0	13.0	102.9	53.7	30
1986	4	4	.0	2446524.5	17	37.859	-42	54.83	17	40.460	-42	55.96	.48	-26.92	1.22	26.76	4.0	12.9	106.7	51.5	46
1986	4	5	.0	2446525.5	17	19.738	-44	11.82	17	22.367	-44	13.89	.46	-23.90	1.24	26.74	4.0	12.9	110.6	49.1	61
1986	4	6	.0	2446526.5	16	59.592	-45	21.01	17	2.240	-45	24.11	.45	-20.55	1.26	26.72	4.0	12.8	114.7	46.4	77
1986	4	7	.0	2446527.5	16	37.445	-46	18.61	16	40.098	-46	22.83	.44	-16.87	1.27	26.69	3.9	12.8	118.8	43.6	93
1986	4	8	.0	2446528.5	16	13.490	-47	.62	16	16.119	-47	6.01	.43	-12.87	1.29	26.66	3.9	12.8	123.1	40.7	108
1986	4	9	.0	2446529.5	15	48.072	-47	23.32	15	50.676	-47	29.88	.42	-8.61	1.30	26.63	3.9	12.8	127.3	37.7	123
1986	4	10	.0	2446530.5	15	21.785	-47	23.82	15	24.332	-47	31.52	.42	-4.14	1.32	26.59	4.0	12.8	131.5	34.8	137
1986	4	11	.0	2446531.5	14	55.313	-47	.73	14	57.785	-47	9.47	.42	.45	1.33	26.55	4.0	12.8	135.5	31.8	147
1986	4	12	.0	2446532.5	14	29.382	-46	14.43	14	31.767	-46	24.08	.42	5.08	1.35	26.51	4.0	12.9	139.2	29.1	152
1986	4	13	.0	2446533.5	14	4.639	-45	7.06	14	6.931	-45	17.48	.42	9.65	1.36	26.47	4.1	12.9	142.4	26.7	148
1986	4	14	.0	2446534.5	13	41.560	-43	42.13	13	43.765	-43	53.15	.43	14.08	1.38	26.43	4.2	13.0	145.2	24.6	137
1986	4	15	.0	2446535.5	13	20.433	-42	3.87	13	22.556	-42	15.33	.44	18.31	1.39	26.38	4.3	13.0	147.3	22.9	125
1986	4	16	.0	2446536.5	13	1.359	-40	16.62	13	3.412	-40	28.40	.45	22.27	1.41	26.34	4.4	13.1	148.7	21.7	111
1986	4	17	.0	2446537.5	12	44.308	-38	24.38	12	46.302	-38	36.38	.47	25.94	1.42	26.29	4.5	13.2	149.3	21.1	97
1986	4	18	.0	2446538.5	12	29.158	-36	30.52	12	31.105	-36	42.66	.48	29.29	1.44	26.24	4.6	13.3	149.3	20.9	83
1986	4	19	.0	2446539.5	12	15.744	-34	37.68	12	17.651	-34	49.90	.50	32.34	1.45	26.19	4.7	13.4	148.7	21.0	69
1986	4	20	.0	2446540.5	12	3.880	-32	47.80	12	5.757	-33	.05	.52	35.07	1.47	26.14	4.8	13.5	147.7	21.4	56
1986	4	21	.0	2446541.5	11	53.386	-31	2.20	11	55.239	-31	14.45	.54	37.53	1.48	26.09	4.9	13.6	146.4	22.0	43
1986	4	22	.0	2446542.5	11	44.091	-29	21.71	11	45.926	-29	33.93	.56	39.71	1.50	26.04	5.0	13.7	144.8	22.8	33
1986	4	23	.0	2446543.5	11	35.843	-27	46.77	11	37.664	-27	58.95	.59	41.66	1.51	25.99	5.2	13.8	143.0	23.5	29
1986	4	24	.0	2446544.5	11	28.508	-26	17.53	11	30.318	-26	29.67	.61	43.39	1.53	25.93	5.3	13.9	141.2	24.3	34
1986	4	25	.0	2446545.5	11	21.967	-24	53.98	11	23.770	-25	6.06	.64	44.94	1.54	25.88	5.4	14.1	139.3	25.1	44
1986	4	26	.0	2446546.5	11	16.121	-23	35.94	11	17.918	-23	47.97	.66	46.31	1.56	25.83	5.5	14.2	137.5	25.9	57
1986	4	27	.0	2446547.5	11	10.883	-22	23.17	11	12.676	-22	35.14	.69	47.52	1.57	25.77	5.6	14.3	135.6	26.6	72
1986	4	28	.0	2446548.5	11	6.178	-21	15.39	11	7.968	-21	27.30	.72	48.61	1.59	25.72	5.8	14.4	133.8	27.2	86
1986	4	29	.0	2446549.5	11	1.943	-20	12.27	11	3.731	-20	24.13	.74	49.58	1.60	25.66	5.9	14.5	132.0	27.8	100
1986	4	30	.0	2446550.5	10	58.123	-19	13.50	10	59.910	-19	25.30	.77	50.45	1.62	25.61	6.0	14.6	130.2	28.4	114
1986	5	1	.0	2446551.5	10	54.671	-18	18.77	10	56.457	-18	30.52	.80	51.22	1.63	25.55	6.1	14.7	128.5	28.9	127
1986	5	2	.0	2446552.5	10	51.546	-17	27.78	10	53.333	-17	39.48	.83	51.92	1.65	25.49	6.2	14.8	126.9	29.3	139
1986	5	3	.0	2446553.5	10	48.715	-16	40.24	10	50.502	-16	51.89	.86	52.54	1.66	25.44	6.3	14.9	125.2	29.7	148
1986	5	4	.0	2446554.5	10	46.145	-15	55.89	10	47.933	-16	7.49	.89	53.10	1.68	25.38	6.4	15.0	123.7	30.0	154
1986	5	5	.0	2446555.5	10	43.812	-15	14.48	10	45.600	-15	26.04	.92	53.60	1.69	25.32	6.5	15.1	122.1	30.3	153
1986	5	6	.0	2446556.5	10	41.691	-14	35.79	10	43.480	-14	47.31	.96	54.05	1.71	25.27	6.6	15.2	120.6	30.6	147
1986	5	7	.0	2446557.5	10	39.762	-13	59.61	10	41.552	-14	11.09	.99	54.45	1.72	25.21	6.7	15.3	119.2	30.8	138
1986	5	8	.0	2446558.5	10	38.008	-13	25.74	10	39.799	-13	37.19	1.02	54.81	1.74	25.15	6.8	15.3	117.8	31.0	127
1986	5	9	.0	2446559.5	10	36.413	-12	54.03	10	38.205	-13	5.44	1.05	55.13	1.75	25.10	6.9	15.4	116.4	31.1	116
1986	5	10	.0	2446560.5	10	34.962	-12	24.29	10	36.756	-12	35.67	1.08	55.41	1.76	25.04	7.0	15.5	115.0	31.2	105
1986	5	11	.0	2446561.5	10	33.644	-11	56.39	10	35.439	-12	7.75	1.11	55.66	1.78	24.98	7.1	15.6	113.7	31.3	94
1986	5	12	.0	2446562.5	10	32.447	-11	30.20	10	34.243	-11	41.53	1.15	55.89	1.79	24.92	7.2	15.7	112.4	31.4	82
1986	5	13	.0	2446563.5	10	31.361	-11	5.60	10	33.159	-11	16.90	1.18	56.09	1.81	24.87	7.3	15.7	111.1	31.4	71
1986	5	14	.0	2446564.5	10	30.378	-10	42.46	10	32.177	-10	53.75	1.21	56.26	1.82	24.81	7.4	15.8	109.9	31.4	60
1986	5	15	.0	2446565.5	10	29.490	-10	20.70	10	31.290	-10	31.97	1.24	56.41	1.84	24.76	7.4	15.9	108.7	31.4	49
1986	5	16	.0	2446566.5	10	28.688	-10	.22	10	30.490	-10	11.46	1.28	56.54	1.85	24.70	7.5	16.0	107.5	31.4	38
1986	5	17	.0	2446567.5	10	27.968	- 9	40.93	10	29.771	- 9	52.15	1.31	56.65	1.87	24.64	7.6	16.0	106.3	31.4	28
1986	5	18	.0	2446568.5	10	27.323	- 9	22.75	10	29.127	- 9	33.96	1.34	56.74	1.88	24.59	7.7	16.1	105.1	31.3	22
1986	5	19	.0	2446569.5	10	26.748	- 9	5.62	10	28.553	- 9	16.81	1.37	56.81	1.89	24.53	7.8	16.2	104.0	31.2	22

YR	MN	DY	HR	J.D.	R.A. (1950)	.0 DEC.	R.A. APPN	DEC.	DELTA	DELDOT	R	RDOT	TMAG	NMAG	THETA	BETA	MOON
1986	5	20	.0	2446570.5	10 26.237	- 8 49.46	10 28.043	- 9 .64	1.41	56.87	1.91	24.48	7.8	16.2	102.9	31.1	29
1986	5	21	.0	2446571.5	10 25.787	- 8 34.22	10 27.594	- 8 45.39	1.44	56.91	1.92	24.42	7.9	16.3	101.8	31.0	40
1986	5	22	.0	2446572.5	10 25.393	- 8 19.84	10 27.201	- 8 31.00	1.47	56.94	1.94	24.36	8.0	16.4	100.7	30.9	53
1986	5	23	.0	2446573.5	10 25.053	- 8 6.26	10 26.861	- 8 17.41	1.51	56.95	1.95	24.31	8.1	16.4	99.6	30.8	67
1986	5	24	.0	2446574.5	10 24.761	- 7 53.45	10 26.571	- 8 4.59	1.54	56.96	1.96	24.25	8.1	16.5	98.6	30.7	81
1986	5	25	.0	2446575.5	10 24.516	- 7 41.35	10 26.327	- 7 52.49	1.57	56.95	1.98	24.20	8.2	16.6	97.5	30.5	96
1986	5	26	.0	2446576.5	10 24.314	- 7 29.93	10 26.126	- 7 41.06	1.60	56.93	1.99	24.15	8.3	16.6	96.5	30.3	110
1986	5	27	.0	2446577.5	10 24.154	- 7 19.15	10 25.966	- 7 30.27	1.64	56.90	2.01	24.09	8.3	16.7	95.5	30.2	124
1986	5	28	.0	2446578.5	10 24.032	- 7 8.97	10 25.845	- 7 20.09	1.67	56.87	2.02	24.04	8.4	16.7	94.5	30.0	137
1986	5	29	.0	2446579.5	10 23.946	- 6 59.36	10 25.760	- 7 10.48	1.70	56.82	2.03	23.98	8.5	16.8	93.5	29.8	149
1986	5	30	.0	2446580.5	10 23.895	- 6 50.29	10 25.710	- 7 1.41	1.74	56.76	2.05	23.93	8.5	16.9	92.5	29.6	157
1986	5	31	.0	2446581.5	10 23.876	- 6 41.74	10 25.692	- 6 52.85	1.77	56.69	2.06	23.88	8.6	16.9	91.5	29.5	160
1986	6	1	.0	2446582.5	10 23.889	- 6 33.68	10 25.705	- 6 44.79	1.80	56.62	2.08	23.82	8.7	17.0	90.5	29.3	155
1986	6	2	.0	2446583.5	10 23.930	- 6 26.08	10 25.747	- 6 37.19	1.83	56.53	2.09	23.77	8.7	17.0	89.6	29.0	146
1986	6	3	.0	2446584.5	10 24.000	- 6 18.93	10 25.817	- 6 30.04	1.87	56.44	2.10	23.72	8.8	17.1	88.6	28.8	136
1986	6	4	.0	2446585.5	10 24.096	- 6 12.20	10 25.913	- 6 23.31	1.90	56.34	2.12	23.67	8.8	17.1	87.7	28.6	125
1986	6	5	.0	2446586.5	10 24.216	- 6 5.87	10 26.034	- 6 16.98	1.93	56.22	2.13	23.62	8.9	17.2	86.7	28.4	114
1986	6	6	.0	2446587.5	10 24.361	- 5 59.93	10 26.179	- 6 11.04	1.96	56.11	2.14	23.56	8.9	17.2	85.8	28.2	103
1986	6	7	.0	2446588.5	10 24.529	- 5 54.35	10 26.347	- 6 5.46	2.00	55.98	2.16	23.51	9.0	17.3	84.9	27.9	92
1986	6	8	.0	2446589.5	10 24.717	- 5 49.12	10 26.536	- 6 .24	2.03	55.84	2.17	23.46	9.1	17.3	83.9	27.7	81
1986	6	9	.0	2446590.5	10 24.926	- 5 44.24	10 26.746	- 5 55.36	2.06	55.70	2.18	23.41	9.1	17.4	83.0	27.5	69
1986	6	10	.0	2446591.5	10 25.154	- 5 39.67	10 26.975	- 5 50.79	2.09	55.55	2.20	23.36	9.2	17.4	82.1	27.2	58
1986	6	11	.0	2446592.5	10 25.401	- 5 35.41	10 27.222	- 5 46.54	2.13	55.39	2.21	23.31	9.2	17.5	81.2	27.0	47
1986	6	12	.0	2446593.5	10 25.666	- 5 31.45	10 27.497	- 5 42.58	2.16	55.23	2.22	23.26	9.3	17.5	80.3	26.7	36
1986	6	13	.0	2446594.5	10 25.947	- 5 27.77	10 27.769	- 5 38.91	2.19	55.06	2.24	23.21	9.3	17.6	79.4	26.5	26
1986	6	14	.0	2446595.5	10 26.245	- 5 24.36	10 28.066	- 5 35.50	2.22	54.88	2.25	23.16	9.4	17.6	78.6	26.2	19
1986	6	15	.0	2446596.5	10 26.557	- 5 21.22	10 28.379	- 5 32.36	2.25	54.69	2.26	23.11	9.4	17.6	77.7	26.0	18
1986	6	16	.0	2446597.5	10 26.884	- 5 18.32	10 28.706	- 5 29.47	2.28	54.50	2.28	23.06	9.5	17.7	76.8	25.7	25
1986	6	17	.0	2446598.5	10 27.225	- 5 15.67	10 29.047	- 5 26.82	2.32	54.31	2.29	23.02	9.5	17.7	76.0	25.5	35
1986	6	18	.0	2446599.5	10 27.579	- 5 13.24	10 29.401	- 5 24.40	2.35	54.10	2.30	22.97	9.6	17.8	75.1	25.2	48
1986	6	19	.0	2446600.5	10 27.946	- 5 11.04	10 29.768	- 5 22.21	2.38	53.90	2.32	22.92	9.6	17.8	74.2	25.0	61
1986	6	20	.0	2446601.5	10 28.325	- 5 9.05	10 30.147	- 5 20.23	2.41	53.68	2.33	22.87	9.7	17.8	73.4	24.7	75
1986	6	21	.0	2446602.5	10 28.715	- 5 7.27	10 30.538	- 5 18.46	2.44	53.46	2.34	22.82	9.7	17.9	72.5	24.4	90
1986	6	22	.0	2446603.5	10 29.116	- 5 5.69	10 30.939	- 5 16.88	2.47	53.24	2.36	22.78	9.8	17.9	71.7	24.2	104
1986	6	23	.0	2446604.5	10 29.527	- 5 4.30	10 31.351	- 5 15.50	2.50	53.01	2.37	22.73	9.8	18.0	70.8	23.9	119
1986	6	24	.0	2446605.5	10 29.949	- 5 3.10	10 31.773	- 5 14.30	2.53	52.78	2.38	22.68	9.9	18.0	70.0	23.6	132
1986	6	25	.0	2446606.5	10 30.381	- 5 2.07	10 32.205	- 5 13.28	2.56	52.55	2.40	22.64	9.9	18.0	69.2	23.3	145
1986	6	26	.0	2446607.5	10 30.821	- 5 1.21	10 32.645	- 5 12.43	2.59	52.30	2.41	22.59	9.9	18.1	68.3	23.1	156
1986	6	27	.0	2446608.5	10 31.271	- 5 .53	10 33.095	- 5 11.76	2.62	52.06	2.42	22.55	10.0	18.1	67.5	22.8	162
1986	6	28	.0	2446609.5	10 31.729	- 5 .00	10 33.553	- 5 11.24	2.65	51.80	2.44	22.50	10.0	18.2	66.7	22.5	160
1986	6	29	.0	2446610.5	10 32.196	- 4 59.63	10 34.020	- 5 10.88	2.68	51.55	2.45	22.45	10.1	18.2	65.8	22.3	151
1986	6	30	.0	2446611.5	10 32.670	- 4 59.42	10 34.494	- 5 10.67	2.71	51.28	2.46	22.41	10.1	18.2	65.0	22.0	141
1986	7	1	.0	2446612.5	10 33.152	- 4 59.35	10 34.976	- 5 10.61	2.74	51.02	2.47	22.36	10.2	18.3	64.2	21.7	130
1986	7	2	.0	2446613.5	10 33.641	- 4 59.43	10 35.466	- 5 10.70	2.77	50.74	2.49	22.32	10.2	18.3	63.4	21.4	119
1986	7	3	.0	2446614.5	10 34.138	- 4 59.64	10 35.962	- 5 10.92	2.80	50.46	2.50	22.28	10.2	18.3	62.6	21.2	108
1986	7	4	.0	2446615.5	10 34.641	- 4 59.99	10 36.465	- 5 11.28	2.83	50.18	2.51	22.23	10.3	18.4	61.8	20.9	97
1986	7	5	.0	2446616.5	10 35.150	- 5 .48	10 36.975	- 5 11.77	2.86	49.89	2.53	22.19	10.3	18.4	61.0	20.6	85
1986	7	6	.0	2446617.5	10 35.666	- 5 1.09	10 37.491	- 5 12.39	2.89	49.60	2.54	22.14	10.4	18.4	60.2	20.3	74
1986	7	7	.0	2446618.5	10 36.188	- 5 1.82	10 38.012	- 5 13.13	2.92	49.30	2.55	22.10	10.4	18.5	59.4	20.0	63
1986	7	8	.0	2446619.5	10 36.715	- 5 2.68	10 38.539	- 5 14.00	2.94	49.00	2.56	22.06	10.4	18.5	58.6	19.8	52

YR	MN	DY	HR	J.D.	R.A. (1950) .0	DEC.	R.A. APPN	DEC.	DELTA	DELDOT	R	RDOT	TMAG	NMAG	THETA	BETA	MOON
1986	7	9	.0	2446620.5	10 37.247	- 5 3.65	10 39.072	- 5 14.98	2.97	48.69	2.58	22.02	10.5	18.5	57.8	19.5	40
1986	7	10	.0	2446621.5	10 37.785	- 5 4.73	10 39.610	- 5 16.07	3.00	48.37	2.59	21.97	10.5	18.6	57.0	19.2	30
1986	7	11	.0	2446622.5	10 38.327	- 5 5.93	10 40.152	- 5 17.27	3.03	48.06	2.60	21.93	10.5	18.6	56.2	18.9	20
1986	7	12	.0	2446623.5	10 38.874	- 5 7.23	10 40.699	- 5 18.59	3.06	47.73	2.62	21.89	10.6	18.6	55.4	18.7	15
1986	7	13	.0	2446624.5	10 39.426	- 5 8.64	10 41.250	- 5 20.00	3.08	47.41	2.63	21.85	10.6	18.6	54.6	18.4	19
1986	7	14	.0	2446625.5	10 39.981	- 5 10.15	10 41.806	- 5 21.52	3.11	47.08	2.64	21.81	10.6	18.7	53.8	18.1	29
1986	7	15	.0	2446626.5	10 40.540	- 5 11.75	10 42.365	- 5 23.13	3.14	46.74	2.65	21.77	10.7	18.7	53.0	17.8	41
1986	7	16	.0	2446627.5	10 41.104	- 5 13.46	10 42.928	- 5 24.85	3.16	46.40	2.67	21.72	10.7	18.7	52.2	17.5	54
1986	7	17	.0	2446628.5	10 41.670	- 5 15.25	10 43.495	- 5 26.65	3.19	46.06	2.68	21.68	10.7	18.8	51.4	17.3	67
1986	7	18	.0	2446629.5	10 42.240	- 5 17.14	10 44.065	- 5 28.55	3.22	45.72	2.69	21.64	10.8	18.8	50.7	17.0	81
1986	7	19	.0	2446630.5	10 42.813	- 5 19.11	10 44.638	- 5 30.53	3.24	45.37	2.70	21.60	10.8	18.8	49.9	16.7	95
1986	7	20	.0	2446631.5	10 43.389	- 5 21.17	10 45.214	- 5 32.60	3.27	45.02	2.72	21.56	10.8	18.8	49.1	16.4	109
1986	7	21	.0	2446632.5	10 43.968	- 5 23.31	10 45.793	- 5 34.75	3.30	44.67	2.73	21.52	10.9	18.9	48.3	16.2	123
1986	7	22	.0	2446633.5	10 44.550	- 5 25.53	10 46.375	- 5 36.98	3.32	44.31	2.74	21.48	10.9	18.9	47.6	15.9	137
1986	7	23	.0	2446634.5	10 45.134	- 5 27.84	10 46.959	- 5 39.29	3.35	43.95	2.75	21.44	10.9	18.9	46.8	15.6	150
1986	7	24	.0	2446635.5	10 45.720	- 5 30.22	10 47.545	- 5 41.68	3.37	43.58	2.77	21.40	11.0	18.9	46.0	15.3	160
1986	7	25	.0	2446636.5	10 46.305	- 5 32.67	10 48.134	- 5 44.15	3.40	43.22	2.78	21.37	11.0	19.0	45.3	15.1	164
1986	7	26	.0	2446637.5	10 46.900	- 5 35.20	10 48.724	- 5 46.68	3.42	42.85	2.79	21.33	11.0	19.0	44.5	14.8	158
1986	7	27	.0	2446638.5	10 47.493	- 5 37.80	10 49.317	- 5 49.29	3.45	42.47	2.80	21.29	11.1	19.0	43.7	14.5	148
1986	7	28	.0	2446639.5	10 48.088	- 5 40.47	10 49.912	- 5 51.97	3.47	42.09	2.81	21.25	11.1	19.0	43.0	14.2	137
1986	7	29	.0	2446640.5	10 48.684	- 5 43.22	10 50.509	- 5 54.72	3.50	41.71	2.83	21.21	11.1	19.1	42.2	14.0	126
1986	7	30	.0	2446641.5	10 49.282	- 5 46.03	10 51.107	- 5 57.54	3.52	41.32	2.84	21.17	11.2	19.1	41.4	13.7	115
1986	7	31	.0	2446642.5	10 49.882	- 5 48.90	10 51.707	- 6 .43	3.54	40.93	2.85	21.14	11.2	19.1	40.7	13.4	104
1986	8	1	.0	2446643.5	10 50.483	- 5 51.84	10 52.308	- 6 3.38	3.57	40.54	2.86	21.10	11.2	19.1	39.9	13.1	92
1986	8	2	.0	2446644.5	10 51.086	- 5 54.85	10 52.910	- 6 6.39	3.59	40.14	2.88	21.06	11.2	19.2	39.2	12.9	81
1986	8	3	.0	2446645.5	10 51.689	- 5 57.91	10 53.514	- 6 9.47	3.61	39.74	2.89	21.03	11.3	19.2	38.4	12.6	70
1986	8	4	.0	2446646.5	10 52.294	- 6 1.04	10 54.119	- 6 12.60	3.64	39.34	2.90	20.99	11.3	19.2	37.7	12.3	58
1986	8	5	.0	2446647.5	10 52.900	- 6 4.22	10 54.724	- 6 15.80	3.66	38.93	2.91	20.95	11.3	19.2	36.9	12.1	47
1986	8	6	.0	2446648.5	10 53.506	- 6 7.47	10 55.331	- 6 19.05	3.68	38.52	2.92	20.92	11.3	19.3	36.2	11.8	36
1986	8	7	.0	2446649.5	10 54.113	- 6 10.77	10 55.937	- 6 22.36	3.70	38.10	2.94	20.88	11.4	19.3	35.4	11.6	25
1986	8	8	.0	2446650.5	10 54.720	- 6 14.12	10 56.545	- 6 25.72	3.73	37.69	2.95	20.84	11.4	19.3	34.7	11.3	17
1986	8	9	.0	2446651.5	10 55.328	- 6 17.53	10 57.152	- 6 29.14	3.75	37.27	2.96	20.81	11.4	19.3	34.0	11.0	15
1986	8	10	.0	2446652.5	10 55.936	- 6 20.99	10 57.760	- 6 32.60	3.77	36.84	2.97	20.77	11.5	19.3	33.2	10.8	23
1986	8	11	.0	2446653.5	10 56.544	- 6 24.51	10 58.368	- 6 36.12	3.79	36.42	2.98	20.74	11.5	19.4	32.5	10.5	34
1986	8	12	.0	2446654.5	10 57.152	- 6 28.07	10 58.976	- 6 39.69	3.81	35.99	3.00	20.70	11.5	19.4	31.8	10.3	46
1986	8	13	.0	2446655.5	10 57.760	- 6 31.68	10 59.584	- 6 43.31	3.83	35.56	3.01	20.67	11.5	19.4	31.0	10.0	59
1986	8	14	.0	2446656.5	10 58.367	- 6 35.34	11 .192	- 6 46.98	3.85	35.13	3.02	20.63	11.6	19.4	30.3	9.7	73
1986	8	15	.0	2446657.5	10 58.974	- 6 39.04	11 .799	- 6 50.69	3.87	34.69	3.03	20.60	11.6	19.4	29.6	9.5	86
1986	8	16	.0	2446658.5	10 59.581	- 6 42.79	11 1.406	- 6 54.45	3.89	34.26	3.04	20.56	11.6	19.5	28.9	9.2	100
1986	8	17	.0	2446659.5	11 .187	- 6 46.58	11 2.012	- 6 58.25	3.91	33.82	3.06	20.53	11.6	19.5	28.2	9.0	114
1986	8	18	.0	2446660.5	11 .793	- 6 50.41	11 2.618	- 7 2.09	3.93	33.38	3.07	20.50	11.7	19.5	27.4	8.7	128
1986	8	19	.0	2446661.5	11 1.398	- 6 54.29	11 3.223	- 7 5.97	3.95	32.94	3.08	20.46	11.7	19.5	26.7	8.5	141
1986	8	20	.0	2446662.5	11 2.001	- 6 58.20	11 3.827	- 7 9.89	3.97	32.49	3.09	20.43	11.7	19.5	26.0	8.3	153
1986	8	21	.0	2446663.5	11 2.604	- 7 2.16	11 4.429	- 7 13.86	3.99	32.05	3.10	20.39	11.7	19.6	25.3	8.0	163
1986	8	22	.0	2446664.5	11 3.206	- 7 6.15	11 5.031	- 7 17.86	4.01	31.60	3.12	20.36	11.7	19.6	24.7	7.8	164
1986	8	23	.0	2446665.5	11 3.807	- 7 10.18	11 5.632	- 7 21.90	4.02	31.15	3.13	20.33	11.8	19.6	24.0	7.6	156
1986	8	24	.0	2446666.5	11 4.407	- 7 14.25	11 6.232	- 7 25.97	4.04	30.70	3.14	20.29	11.8	19.6	23.3	7.3	146
1986	8	25	.0	2446667.5	11 5.005	- 7 18.36	11 6.830	- 7 30.08	4.06	30.25	3.15	20.26	11.8	19.6	22.6	7.1	134
1986	8	26	.0	2446668.5	11 5.603	- 7 22.50	11 7.428	- 7 34.23	4.08	29.79	3.16	20.23	11.8	19.7	22.0	6.9	123
1986	8	27	.0	2446669.5	11 6.198	- 7 26.68	11 8.024	- 7 38.42	4.09	29.33	3.17	20.20	11.9	19.7	21.3	6.6	112

YR	MN	DY	HR	J.D.	R.A. (1950).0	DEC.	R.A. APPN	DEC.	DELTA	DELDOT	R	RDOT	TMAG	NMAG	THETA	BETA	MOON
1986	8	28	.0	2446670.5	11 6.793	- 7 30.89	11 8.618	- 7 42.64	4.11	28.87	3.19	20.16	11.9	19.7	20.7	6.4	100
1986	8	29	.0	2446671.5	11 7.385	- 7 35.14	11 9.211	- 7 46.89	4.13	28.41	3.20	20.13	11.9	19.7	20.0	6.2	89
1986	8	30	.0	2446672.5	11 7.976	- 7 39.42	11 9.802	- 7 51.18	4.14	27.94	3.21	20.10	11.9	19.7	19.4	6.0	78
1986	8	31	.0	2446673.5	11 8.565	- 7 43.73	11 10.391	- 7 55.49	4.16	27.48	3.22	20.07	11.9	19.7	18.8	5.8	66
1986	9	1	.0	2446674.5	11 9.152	- 7 48.07	11 10.978	- 7 59.84	4.17	27.01	3.23	20.04	12.0	19.8	18.2	5.6	55
1986	9	2	.0	2446675.5	11 9.738	- 7 52.44	11 11.564	- 8 4.22	4.19	26.54	3.24	20.01	12.0	19.8	17.6	5.4	43
1986	9	3	.0	2446676.5	11 10.321	- 7 56.84	11 12.147	- 8 8.63	4.21	26.06	3.25	19.97	12.0	19.8	17.1	5.2	32
1986	9	4	.0	2446677.5	11 10.902	- 8 1.28	11 12.728	- 8 13.07	4.22	25.59	3.27	19.94	12.0	19.8	16.5	5.0	21
1986	9	5	.0	2446678.5	11 11.480	- 8 5.73	11 13.306	- 8 17.53	4.24	25.11	3.28	19.91	12.0	19.8	16.0	4.9	15
1986	9	6	.0	2446679.5	11 12.056	- 8 10.22	11 13.882	- 8 22.02	4.25	24.64	3.29	19.88	12.1	19.8	15.5	4.7	17
1986	9	7	.0	2446680.5	11 12.629	- 8 14.73	11 14.456	- 8 26.54	4.26	24.16	3.30	19.85	12.1	19.8	15.1	4.6	27
1986	9	8	.0	2446681.5	11 13.200	- 8 19.27	11 15.027	- 8 31.08	4.28	23.68	3.31	19.82	12.1	19.9	14.6	4.4	39
1986	9	9	.0	2446682.5	11 13.768	- 8 23.83	11 15.595	- 8 35.65	4.29	23.20	3.32	19.79	12.1	19.9	14.2	4.3	52
1986	9	10	.0	2446683.5	11 14.333	- 8 28.42	11 16.160	- 8 40.24	4.30	22.72	3.34	19.76	12.1	19.9	13.8	4.1	65
1986	9	11	.0	2446684.5	11 14.895	- 8 33.02	11 16.722	- 8 44.86	4.32	22.24	3.35	19.73	12.1	19.9	13.5	4.0	79
1986	9	12	.0	2446685.5	11 15.454	- 8 37.65	11 17.281	- 8 49.49	4.33	21.76	3.36	19.70	12.2	19.9	13.2	3.9	92
1986	9	13	.0	2446686.5	11 16.010	- 8 42.30	11 17.837	- 8 54.15	4.34	21.28	3.37	19.67	12.2	19.9	13.0	3.8	106
1986	9	14	.0	2446687.5	11 16.562	- 8 46.97	11 18.390	- 8 58.82	4.35	20.80	3.38	19.64	12.2	19.9	12.8	3.8	119
1986	9	15	.0	2446688.5	11 17.111	- 8 51.66	11 18.939	- 9 3.52	4.37	20.32	3.39	19.61	12.2	20.0	12.6	3.7	132
1986	9	16	.0	2446689.5	11 17.656	- 8 56.37	11 19.485	- 9 8.23	4.38	19.84	3.40	19.58	12.2	20.0	12.5	3.7	145
1986	9	17	.0	2446690.5	11 18.198	- 9 1.10	11 20.027	- 9 12.96	4.39	19.36	3.41	19.55	12.2	20.0	12.4	3.6	156
1986	9	18	.0	2446691.5	11 18.736	- 9 5.84	11 20.565	- 9 17.71	4.40	18.87	3.43	19.52	12.3	20.0	12.4	3.6	164
1986	9	19	.0	2446692.5	11 19.271	- 9 10.60	11 21.100	- 9 22.48	4.41	18.39	3.44	19.50	12.3	20.0	12.5	3.6	163
1986	9	20	.0	2446693.5	11 19.801	- 9 15.38	11 21.630	- 9 27.26	4.42	17.91	3.45	19.47	12.3	20.0	12.6	3.6	154
1986	9	21	.0	2446694.5	11 20.328	- 9 20.18	11 22.157	- 9 32.06	4.43	17.43	3.46	19.44	12.3	20.0	12.8	3.7	143
1986	9	22	.0	2446695.5	11 20.851	- 9 24.98	11 22.680	- 9 36.87	4.44	16.95	3.47	19.41	12.3	20.0	13.0	3.7	131
1986	9	23	.0	2446696.5	11 21.369	- 9 29.81	11 23.199	- 9 41.70	4.45	16.47	3.48	19.38	12.3	20.1	13.2	3.8	120
1986	9	24	.0	2446697.5	11 21.883	- 9 34.65	11 23.713	- 9 46.54	4.46	15.98	3.49	19.35	12.4	20.1	13.5	3.9	109
1986	9	25	.0	2446698.5	11 22.393	- 9 39.50	11 24.224	- 9 51.40	4.47	15.50	3.50	19.33	12.4	20.1	13.9	3.9	97
1986	9	26	.0	2446699.5	11 22.899	- 9 44.36	11 24.730	- 9 56.27	4.48	15.02	3.52	19.30	12.4	20.1	14.3	4.0	86
1986	9	27	.0	2446700.5	11 23.400	- 9 49.24	11 25.231	-10 1.15	4.49	14.53	3.53	19.27	12.4	20.1	14.7	4.1	75
1986	9	28	.0	2446701.5	11 23.896	- 9 54.13	11 25.728	-10 6.04	4.50	14.05	3.54	19.24	12.4	20.1	15.2	4.2	63
1986	9	29	.0	2446702.5	11 24.388	- 9 59.03	11 26.220	-10 10.95	4.50	13.56	3.55	19.21	12.4	20.1	15.6	4.4	52
1986	9	30	.0	2446703.5	11 24.875	-10 3.94	11 26.707	-10 15.86	4.51	13.08	3.56	19.19	12.4	20.1	16.2	4.5	40
1986	10	1	.0	2446704.5	11 25.356	-10 8.86	11 27.189	-10 20.79	4.52	12.60	3.57	19.16	12.5	20.1	16.7	4.6	29
1986	10	2	.0	2446705.5	11 25.833	-10 13.79	11 27.665	-10 25.72	4.53	12.12	3.58	19.13	12.5	20.1	17.3	4.8	19
1986	10	3	.0	2446706.5	11 26.304	-10 18.72	11 28.137	-10 30.66	4.53	11.63	3.59	19.11	12.5	20.2	17.9	4.9	14
1986	10	4	.0	2446707.5	11 26.770	-10 23.67	11 28.603	-10 35.61	4.54	11.15	3.60	19.08	12.5	20.2	18.5	5.0	20
1986	10	5	.0	2446708.5	11 27.230	-10 28.62	11 29.064	-10 40.56	4.55	10.67	3.62	19.05	12.5	20.2	19.1	5.2	31
1986	10	6	.0	2446709.5	11 27.685	-10 33.57	11 29.519	-10 45.52	4.55	10.20	3.63	19.03	12.5	20.2	19.8	5.3	44
1986	10	7	.0	2446710.5	11 28.134	-10 38.53	11 29.968	-10 50.48	4.56	9.72	3.64	19.00	12.5	20.2	20.4	5.5	58
1986	10	8	.0	2446711.5	11 28.577	-10 43.50	11 30.411	-10 55.45	4.56	9.24	3.65	18.97	12.6	20.2	21.1	5.7	71
1986	10	9	.0	2446712.5	11 29.014	-10 48.46	11 30.848	-11 .42	4.57	8.77	3.66	18.95	12.6	20.2	21.8	5.8	85
1986	10	10	.0	2446713.5	11 29.444	-10 53.43	11 31.279	-11 5.40	4.57	8.30	3.67	18.92	12.6	20.2	22.5	6.0	99
1986	10	11	.0	2446714.5	11 29.868	-10 58.40	11 31.704	-11 10.37	4.58	7.83	3.68	18.89	12.6	20.2	23.2	6.1	112
1986	10	12	.0	2446715.5	11 30.286	-11 3.38	11 32.122	-11 15.35	4.58	7.37	3.69	18.87	12.6	20.2	23.9	6.3	125
1986	10	13	.0	2446716.5	11 30.697	-11 8.35	11 32.534	-11 20.32	4.59	6.91	3.70	18.84	12.6	20.2	24.7	6.5	138
1986	10	14	.0	2446717.5	11 31.102	-11 13.32	11 32.939	-11 25.29	4.59	6.45	3.71	18.82	12.6	20.3	25.4	6.6	150
1986	10	15	.0	2446718.5	11 31.500	-11 18.28	11 33.337	-11 30.26	4.59	5.99	3.72	18.79	12.6	20.3	26.2	6.8	160
1986	10	16	.0	2446719.5	11 31.891	-11 23.25	11 33.728	-11 35.23	4.60	5.53	3.74	18.77	12.6	20.3	26.9	6.9	164

YR	MN	DY	HR	J.D.	R.A. (1950) .0 DEC.	R.A. APPN DEC.	DELTA	DELDOT	R	RDOT	TMAG	NMAG	THETA	BETA	MOON
1986	10	17	.0	2446720.5	11 32.274 -11 28.21	11 34.112 -11 40.20	4.60	5.08	3.75	18.74	12.7	20.3	27.7	7.1	159
1986	10	18	.0	2446721.5	11 32.651 -11 33.17	11 34.489 -11 45.16	4.60	4.63	3.76	18.72	12.7	20.3	28.5	7.3	150
1986	10	19	.0	2446722.5	11 33.021 -11 38.13	11 34.859 -11 50.12	4.61	4.18	3.77	18.69	12.7	20.3	29.3	7.4	139
1986	10	20	.0	2446723.5	11 33.383 -11 43.08	11 35.221 -11 55.07	4.61	3.73	3.78	18.67	12.7	20.3	30.1	7.6	127
1986	10	21	.0	2446724.5	11 33.737 -11 48.03	11 35.576 -12 .02	4.61	3.29	3.79	18.64	12.7	20.3	30.9	7.7	116
1986	10	22	.0	2446725.5	11 34.084 -11 52.96	11 35.924 -12 4.97	4.61	2.84	3.80	18.62	12.7	20.3	31.7	7.9	105
1986	10	23	.0	2446726.5	11 34.423 -11 57.90	11 36.263 -12 9.90	4.61	2.40	3.81	18.59	12.7	20.3	32.5	8.1	93
1986	10	24	.0	2446727.5	11 34.754 -12 2.82	11 36.595 -12 14.83	4.61	1.97	3.82	18.57	12.7	20.3	33.3	8.2	82
1986	10	25	.0	2446728.5	11 35.077 -12 7.74	11 36.919 -12 19.75	4.61	1.53	3.83	18.54	12.7	20.3	34.1	8.4	71
1986	10	26	.0	2446729.5	11 35.392 -12 12.65	11 37.234 -12 24.66	4.62	1.10	3.84	18.52	12.8	20.3	34.9	8.5	59
1986	10	27	.0	2446730.5	11 35.699 -12 17.54	11 37.541 -12 29.56	4.62	.67	3.85	18.49	12.8	20.4	35.8	8.7	48
1986	10	28	.0	2446731.5	11 35.997 -12 22.43	11 37.839 -12 34.45	4.62	.24	3.86	18.47	12.8	20.4	36.6	8.8	37
1986	10	29	.0	2446732.5	11 36.286 -12 27.30	11 38.129 -12 39.32	4.62	-.18	3.88	18.45	12.8	20.4	37.4	9.0	26
1986	10	30	.0	2446733.5	11 36.566 -12 32.16	11 38.410 -12 44.19	4.62	-.60	3.89	18.42	12.8	20.4	38.3	9.1	17
1986	10	31	.0	2446734.5	11 36.838 -12 37.01	11 38.681 -12 49.04	4.62	-1.01	3.90	18.40	12.8	20.4	39.1	9.3	16
1986	11	1	.0	2446735.5	11 37.100 -12 41.84	11 38.944 -12 53.87	4.62	-1.42	3.91	18.37	12.8	20.4	40.0	9.4	24
1986	11	2	.0	2446736.5	11 37.352 -12 46.66	11 39.197 -12 58.69	4.61	-1.83	3.92	18.35	12.8	20.4	40.8	9.5	36
1986	11	3	.0	2446737.5	11 37.596 -12 51.46	11 39.441 -13 3.49	4.61	-2.23	3.93	18.33	12.8	20.4	41.7	9.7	50
1986	11	4	.0	2446738.5	11 37.829 -12 56.24	11 39.675 -13 8.27	4.61	-2.63	3.94	18.30	12.8	20.4	42.6	9.8	64
1986	11	5	.0	2446739.5	11 38.053 -13 1.00	11 39.899 -13 13.04	4.61	-3.02	3.95	18.28	12.8	20.4	43.4	9.9	78
1986	11	6	.0	2446740.5	11 38.266 -13 5.74	11 40.113 -13 17.78	4.61	-3.41	3.96	18.26	12.8	20.4	44.3	10.1	92
1986	11	7	.0	2446741.5	11 38.470 -13 10.46	11 40.317 -13 22.50	4.61	-3.79	3.97	18.23	12.9	20.4	45.2	10.2	106
1986	11	8	.0	2446742.5	11 38.663 -13 15.15	11 40.510 -13 27.20	4.60	-4.16	3.98	18.21	12.9	20.4	46.1	10.3	119
1986	11	9	.0	2446743.5	11 38.845 -13 19.82	11 40.693 -13 31.88	4.60	-4.53	3.99	18.19	12.9	20.4	47.0	10.5	132
1986	11	10	.0	2446744.5	11 39.017 -13 24.47	11 40.866 -13 36.53	4.60	-4.90	4.00	18.17	12.9	20.4	47.9	10.6	144
1986	11	11	.0	2446745.5	11 39.178 -13 29.09	11 41.027 -13 41.15	4.60	-5.26	4.01	18.14	12.9	20.4	48.7	10.7	155
1986	11	12	.0	2446746.5	11 39.328 -13 33.69	11 41.177 -13 45.74	4.59	-5.61	4.02	18.12	12.9	20.4	49.6	10.8	162
1986	11	13	.0	2446747.5	11 39.467 -13 38.25	11 41.317 -13 50.31	4.59	-5.96	4.03	18.10	12.9	20.4	50.5	10.9	162
1986	11	14	.0	2446748.5	11 39.594 -13 42.79	11 41.445 -13 54.85	4.59	-6.30	4.04	18.08	12.9	20.4	51.5	11.0	154
1986	11	15	.0	2446749.5	11 39.711 -13 47.29	11 41.561 -13 59.36	4.58	-6.63	4.05	18.05	12.9	20.4	52.4	11.1	144
1986	11	16	.0	2446750.5	11 39.815 -13 51.77	11 41.667 -14 3.84	4.58	-6.96	4.06	18.03	12.9	20.4	53.3	11.2	133
1986	11	17	.0	2446751.5	11 39.908 -13 56.21	11 41.760 -14 8.28	4.57	-7.29	4.08	18.01	12.9	20.5	54.2	11.3	122
1986	11	18	.0	2446752.5	11 39.989 -14 .62	11 41.842 -14 12.70	4.57	-7.60	4.09	17.99	12.9	20.5	55.1	11.4	110
1986	11	19	.0	2446753.5	11 40.058 -14 5.00	11 41.911 -14 17.07	4.57	-7.92	4.10	17.97	12.9	20.5	56.0	11.5	99
1986	11	20	.0	2446754.5	11 40.115 -14 9.34	11 41.969 -14 21.42	4.56	-8.22	4.11	17.94	12.9	20.5	57.0	11.6	88
1986	11	21	.0	2446755.5	11 40.160 -14 13.64	11 42.013 -14 25.72	4.56	-8.52	4.12	17.92	13.0	20.5	57.9	11.7	76
1986	11	22	.0	2446756.5	11 40.192 -14 17.91	11 42.046 -14 29.99	4.55	-8.82	4.13	17.90	13.0	20.5	58.8	11.8	65
1986	11	23	.0	2446757.5	11 40.211 -14 22.13	11 42.065 -14 34.22	4.55	-9.10	4.14	17.88	13.0	20.5	59.8	11.9	54
1986	11	24	.0	2446758.5	11 40.217 -14 26.32	11 42.072 -14 38.40	4.54	-9.38	4.15	17.86	13.0	20.5	60.7	12.0	43
1986	11	25	.0	2446759.5	11 40.210 -14 30.46	11 42.066 -14 42.55	4.53	-9.66	4.16	17.84	13.0	20.5	61.7	12.1	32
1986	11	26	.0	2446760.5	11 40.190 -14 34.56	11 42.046 -14 46.65	4.53	-9.92	4.17	17.81	13.0	20.5	62.6	12.1	22
1986	11	27	.0	2446761.5	11 40.156 -14 38.61	11 42.013 -14 50.70	4.52	-10.18	4.18	17.79	13.0	20.5	63.6	12.2	17
1986	11	28	.0	2446762.5	11 40.109 -14 42.62	11 41.966 -14 54.71	4.52	-10.43	4.19	17.77	13.0	20.5	64.5	12.3	20
1986	11	29	.0	2446763.5	11 40.048 -14 46.57	11 41.905 -14 58.67	4.51	-10.67	4.20	17.75	13.0	20.5	65.5	12.3	30
1986	11	30	.0	2446764.5	11 39.973 -14 50.48	11 41.830 -15 2.58	4.50	-10.91	4.21	17.73	13.0	20.5	66.5	12.4	42
1986	12	1	.0	2446765.5	11 39.884 -14 54.34	11 41.742 -15 6.44	4.50	-11.13	4.22	17.71	13.0	20.5	67.4	12.5	56
1986	12	2	.0	2446766.5	11 39.781 -14 58.14	11 41.639 -15 10.24	4.49	-11.35	4.23	17.69	13.0	20.5	68.4	12.5	70
1986	12	3	.0	2446767.5	11 39.662 -15 1.89	11 41.521 -15 13.99	4.49	-11.56	4.24	17.67	13.0	20.5	69.4	12.6	85
1986	12	4	.0	2446768.5	11 39.530 -15 5.57	11 41.389 -15 17.68	4.48	-11.76	4.25	17.65	13.0	20.5	70.4	12.6	99
1986	12	5	.0	2446769.5	11 39.382 -15 9.20	11 41.242 -15 21.32	4.47	-11.94	4.26	17.63	13.0	20.5	71.4	12.7	113

YR	MN	DY	HR	J.D.	R.A. (1950) .0	DEC.	R.A. APPN	DEC.	DELTA	DELDOT	R	RDOT	TMAG	NMAG	THETA	BETA	MOON
1986	12	6	.0	2446770.5	11 39.220	-15 12.77	11 41.080	-15 24.89	4.46	-12.12	4.27	17.61	13.0	20.5	72.3	12.7	127
1986	12	7	.0	2446771.5	11 39.042	-15 16.28	11 40.902	-15 28.40	4.46	-12.29	4.28	17.58	13.0	20.5	73.3	12.7	140
1986	12	8	.0	2446772.5	11 38.849	-15 19.72	11 40.710	-15 31.84	4.45	-12.45	4.29	17.56	13.0	20.5	74.3	12.8	151
1986	12	9	.0	2446773.5	11 38.641	-15 23.10	11 40.502	-15 35.22	4.44	-12.60	4.30	17.54	13.0	20.5	75.3	12.8	160
1986	12	10	.0	2446774.5	11 38.417	-15 26.41	11 40.278	-15 38.53	4.44	-12.74	4.31	17.52	13.0	20.5	76.3	12.8	161
1986	12	11	.0	2446775.5	11 38.178	-15 29.65	11 40.039	-15 41.77	4.43	-12.87	4.32	17.50	13.1	20.5	77.3	12.8	156
1986	12	12	.0	2446776.5	11 37.923	-15 32.82	11 39.784	-15 44.94	4.42	-12.99	4.33	17.48	13.1	20.5	78.4	12.9	146
1986	12	13	.0	2446777.5	11 37.652	-15 35.92	11 39.514	-15 48.04	4.41	-13.10	4.34	17.46	13.1	20.5	79.4	12.9	135
1986	12	14	.0	2446778.5	11 37.365	-15 38.94	11 39.227	-15 51.06	4.41	-13.20	4.35	17.44	13.1	20.5	80.4	12.9	124
1986	12	15	.0	2446779.5	11 37.063	-15 41.89	11 38.925	-15 54.01	4.40	-13.29	4.36	17.42	13.1	20.5	81.4	12.9	113
1986	12	16	.0	2446780.5	11 36.744	-15 44.75	11 38.606	-15 56.88	4.39	-13.37	4.37	17.40	13.1	20.5	82.4	12.9	102
1986	12	17	.0	2446781.5	11 36.408	-15 47.54	11 38.271	-15 59.67	4.38	-13.44	4.38	17.38	13.1	20.5	83.5	12.9	91
1986	12	18	.0	2446782.5	11 36.057	-15 50.25	11 37.920	-16 2.38	4.38	-13.50	4.39	17.36	13.1	20.5	84.5	12.9	79
1986	12	19	.0	2446783.5	11 35.688	-15 52.88	11 37.552	-16 5.01	4.37	-13.55	4.40	17.35	13.1	20.5	85.5	12.9	68
1986	12	20	.0	2446784.5	11 35.304	-15 55.41	11 37.167	-16 7.55	4.36	-13.59	4.41	17.33	13.1	20.5	86.6	12.9	57
1986	12	21	.0	2446785.5	11 34.902	-15 57.87	11 36.766	-16 10.00	4.35	-13.62	4.42	17.31	13.1	20.5	87.6	12.8	46
1986	12	22	.0	2446786.5	11 34.484	-16 .23	11 36.348	-16 12.36	4.34	-13.64	4.43	17.29	13.1	20.5	88.7	12.8	35
1986	12	23	.0	2446787.5	11 34.049	-16 2.50	11 35.913	-16 14.63	4.34	-13.64	4.44	17.27	13.1	20.5	89.7	12.8	26
1986	12	24	.0	2446788.5	11 33.597	-16 4.67	11 35.461	-16 16.81	4.33	-13.64	4.45	17.25	13.1	20.5	90.8	12.8	19
1986	12	25	.0	2446789.5	11 33.128	-16 6.75	11 34.992	-16 18.89	4.32	-13.62	4.46	17.23	13.1	20.5	91.8	12.7	20
1986	12	26	.0	2446790.5	11 32.642	-16 8.74	11 34.506	-16 20.87	4.31	-13.60	4.47	17.21	13.1	20.5	92.9	12.7	27
1986	12	27	.0	2446791.5	11 32.139	-16 10.62	11 34.003	-16 22.75	4.30	-13.56	4.48	17.19	13.1	20.5	94.0	12.6	38
1986	12	28	.0	2446792.5	11 31.619	-16 12.40	11 33.483	-16 24.53	4.30	-13.50	4.49	17.17	13.1	20.5	95.1	12.6	51
1986	12	29	.0	2446793.5	11 31.082	-16 14.07	11 32.945	-16 26.21	4.29	-13.44	4.50	17.15	13.1	20.5	96.1	12.5	65
1986	12	30	.0	2446794.5	11 30.527	-16 15.64	11 32.391	-16 27.78	4.28	-13.36	4.51	17.14	13.1	20.5	97.2	12.5	79
1986	12	31	.0	2446795.5	11 29.955	-16 17.10	11 31.819	-16 29.23	4.27	-13.27	4.52	17.12	13.1	20.5	98.3	12.4	94
1987	1	1	.0	2446796.5	11 29.366	-16 18.44	11 31.230	-16 30.58	4.27	-13.17	4.53	17.10	13.1	20.5	99.4	12.4	108
1987	1	2	.0	2446797.5	11 28.760	-16 19.67	11 30.624	-16 31.81	4.26	-13.05	4.54	17.08	13.1	20.5	100.5	12.3	123
1987	1	3	.0	2446798.5	11 28.136	-16 20.79	11 30.000	-16 32.93	4.25	-12.92	4.55	17.06	13.1	20.5	101.5	12.2	136
1987	1	4	.0	2446799.5	11 27.496	-16 21.78	11 29.360	-16 33.92	4.24	-12.78	4.56	17.04	13.1	20.5	102.6	12.1	148
1987	1	5	.0	2446800.5	11 26.838	-16 22.66	11 28.702	-16 34.80	4.24	-12.62	4.57	17.02	13.1	20.5	103.7	12.1	157
1987	1	6	.0	2446801.5	11 26.164	-16 23.42	11 28.027	-16 35.55	4.23	-12.45	4.58	17.01	13.1	20.5	104.8	12.0	160
1987	1	7	.0	2446802.5	11 25.473	-16 24.05	11 27.336	-16 36.18	4.22	-12.27	4.59	16.99	13.2	20.5	105.9	11.9	155
1987	1	8	.0	2446803.5	11 24.765	-16 24.55	11 26.628	-16 36.68	4.21	-12.08	4.60	16.97	13.2	20.5	107.0	11.8	146
1987	1	9	.0	2446804.5	11 24.041	-16 24.93	11 25.904	-16 37.06	4.21	-11.87	4.61	16.95	13.2	20.5	108.1	11.7	135
1987	1	10	.0	2446805.5	11 23.300	-16 25.18	11 25.163	-16 37.31	4.20	-11.66	4.62	16.93	13.2	20.5	109.3	11.6	124
1987	1	11	.0	2446806.5	11 22.543	-16 25.30	11 24.406	-16 37.43	4.19	-11.43	4.63	16.92	13.2	20.5	110.4	11.5	113
1987	1	12	.0	2446807.5	11 21.770	-16 25.29	11 23.632	-16 37.41	4.19	-11.18	4.64	16.90	13.2	20.5	111.5	11.4	102
1987	1	13	.0	2446808.5	11 20.981	-16 25.14	11 22.843	-16 37.26	4.18	-10.93	4.65	16.88	13.2	20.5	112.6	11.3	91
1987	1	14	.0	2446809.5	11 20.176	-16 24.86	11 22.038	-16 36.98	4.18	-10.66	4.66	16.86	13.2	20.5	113.7	11.1	79
1987	1	15	.0	2446810.5	11 19.356	-16 24.44	11 21.218	-16 36.55	4.17	-10.38	4.67	16.84	13.2	20.5	114.8	11.0	68
1987	1	16	.0	2446811.5	11 18.520	-16 23.88	11 20.382	-16 35.99	4.16	-10.09	4.68	16.83	13.2	20.5	116.0	10.9	57
1987	1	17	.0	2446812.5	11 17.669	-16 23.19	11 19.530	-16 35.29	4.16	-9.79	4.69	16.81	13.2	20.5	117.1	10.8	46
1987	1	18	.0	2446813.5	11 16.803	-16 22.35	11 18.664	-16 34.45	4.15	-9.47	4.70	16.79	13.2	20.6	118.2	10.6	36
1987	1	19	.0	2446814.5	11 15.923	-16 21.36	11 17.783	-16 33.46	4.15	-9.14	4.71	16.77	13.2	20.6	119.3	10.5	27
1987	1	20	.0	2446815.5	11 15.028	-16 20.24	11 16.888	-16 32.32	4.14	-8.80	4.72	16.76	13.2	20.6	120.4	10.4	21
1987	1	21	.0	2446816.5	11 14.119	-16 18.97	11 15.978	-16 31.05	4.14	-8.45	4.73	16.74	13.2	20.6	121.6	10.2	22
1987	1	22	.0	2446817.5	11 13.196	-16 17.55	11 15.055	-16 29.62	4.13	-8.09	4.74	16.72	13.2	20.6	122.7	10.1	29
1987	1	23	.0	2446818.5	11 12.259	-16 15.98	11 14.118	-16 28.05	4.13	-7.71	4.75	16.70	13.2	20.6	123.8	9.9	39
1987	1	24	.0	2446819.5	11 11.310	-16 14.27	11 13.168	-16 26.33	4.12	-7.32	4.76	16.69	13.2	20.6	124.9	9.8	51

YR	MN	DY	HR	J.D.	R.A. (1950) .0 DEC.			R.A. APPN DEC.			DELTA	DELDOT	R	RDOT	TMAG	NMAG	THETA	BETA	MOON
1987	1	25	.0	2446820.5	11	10.347	-16 12.40	11	12.205	-16 24.45	4.12	-6.92	4.76	16.67	13.2	20.6	126.1	9.6	64
1987	1	26	.0	2446821.5	11	9.372	-16 10.38	11	11.229	-16 22.43	4.11	-6.51	4.77	16.65	13.2	20.6	127.2	9.5	78
1987	1	27	.0	2446822.5	11	8.384	-16 8.22	11	10.242	-16 20.26	4.11	-6.08	4.78	16.64	13.2	20.6	128.3	9.3	92
1987	1	28	.0	2446823.5	11	7.385	-16 5.90	11	9.242	-16 17.93	4.11	-5.64	4.79	16.62	13.2	20.6	129.4	9.1	106
1987	1	29	.0	2446824.5	11	6.374	-16 3.42	11	8.231	-16 15.45	4.10	-5.19	4.80	16.60	13.2	20.6	130.6	9.0	120
1987	1	30	.0	2446825.5	11	5.352	-16 .80	11	7.209	-16 12.82	4.10	-4.73	4.81	16.59	13.2	20.6	131.7	8.8	134
1987	1	31	.0	2446826.5	11	4.320	-15 58.02	11	6.176	-16 10.03	4.10	-4.26	4.82	16.57	13.3	20.6	132.8	8.6	146
1987	2	1	.0	2446827.5	11	3.278	-15 55.09	11	5.133	-16 7.09	4.10	-3.77	4.83	16.55	13.3	20.6	133.9	8.5	155
1987	2	2	.0	2446828.5	11	2.226	-15 52.01	11	4.081	-16 4.00	4.09	-3.28	4.84	16.54	13.3	20.6	135.0	8.3	158
1987	2	3	.0	2446829.5	11	1.165	-15 48.78	11	3.020	-16 .76	4.09	-2.77	4.85	16.52	13.3	20.6	136.1	8.1	152
1987	2	4	.0	2446830.5	11	.096	-15 45.39	11	1.950	-15 57.36	4.09	-2.26	4.86	16.50	13.3	20.6	137.2	7.9	143
1987	2	5	.0	2446831.5	10	59.018	-15 41.86	11	.872	-15 53.82	4.09	-1.74	4.87	16.49	13.3	20.6	138.3	7.7	132
1987	2	6	.0	2446832.5	10	57.933	-15 38.18	10	59.786	-15 50.12	4.09	-1.20	4.88	16.47	13.3	20.6	139.3	7.6	121
1987	2	7	.0	2446833.5	10	56.841	-15 34.35	10	58.694	-15 46.28	4.09	-.66	4.89	16.45	13.3	20.6	140.4	7.4	110
1987	2	8	.0	2446834.5	10	55.743	-15 30.37	10	57.595	-15 42.29	4.09	-.11	4.90	16.44	13.3	20.6	141.5	7.2	99
1987	2	9	.0	2446835.5	10	54.639	-15 26.25	10	56.491	-15 38.16	4.09	.44	4.91	16.42	13.3	20.6	142.5	7.0	88
1987	2	10	.0	2446836.5	10	53.529	-15 21.99	10	55.380	-15 33.88	4.09	1.01	4.92	16.40	13.3	20.6	143.5	6.8	76
1987	2	11	.0	2446837.5	10	52.414	-15 17.58	10	54.265	-15 29.46	4.09	1.58	4.93	16.39	13.3	20.6	144.6	6.7	65
1987	2	12	.0	2446838.5	10	51.295	-15 13.03	10	53.146	-15 24.90	4.09	2.16	4.94	16.37	13.3	20.6	145.6	6.5	54
1987	2	13	.0	2446839.5	10	50.172	-15 8.35	10	52.023	-15 20.20	4.09	2.74	4.95	16.36	13.3	20.6	146.6	6.3	44
1987	2	14	.0	2446840.5	10	49.047	-15 3.53	10	50.897	-15 15.37	4.09	3.33	4.96	16.34	13.3	20.6	147.5	6.1	34
1987	2	15	.0	2446841.5	10	47.918	-14 58.58	10	49.768	-15 10.40	4.10	3.93	4.96	16.32	13.3	20.6	148.5	6.0	26
1987	2	16	.0	2446842.5	10	46.787	-14 53.49	10	48.636	-15 5.30	4.10	4.53	4.97	16.31	13.4	20.6	149.4	5.8	22
1987	2	17	.0	2446843.5	10	45.655	-14 48.28	10	47.504	-15 .07	4.10	5.14	4.98	16.29	13.4	20.7	150.3	5.6	25
1987	2	18	.0	2446844.5	10	44.522	-14 42.93	10	46.370	-14 54.71	4.10	5.76	4.99	16.28	13.4	20.7	151.2	5.5	33
1987	2	19	.0	2446845.5	10	43.388	-14 37.47	10	45.236	-14 49.23	4.11	6.38	5.00	16.26	13.4	20.7	152.0	5.3	44
1987	2	20	.0	2446846.5	10	42.255	-14 31.88	10	44.102	-14 43.62	4.11	7.00	5.01	16.24	13.4	20.7	152.9	5.2	55
1987	2	21	.0	2446847.5	10	41.122	-14 26.17	10	42.969	-14 37.90	4.12	7.63	5.02	16.23	13.4	20.7	153.6	5.0	68
1987	2	22	.0	2446848.5	10	39.990	-14 20.35	10	41.837	-14 32.06	4.12	8.26	5.03	16.21	13.4	20.7	154.4	4.9	81
1987	2	23	.0	2446849.5	10	38.861	-14 14.42	10	40.708	-14 26.11	4.13	8.90	5.04	16.20	13.4	20.7	155.1	4.7	94
1987	2	24	.0	2446850.5	10	37.733	-14 8.37	10	39.580	-14 20.05	4.13	9.54	5.05	16.18	13.4	20.7	155.7	4.6	108
1987	2	25	.0	2446851.5	10	36.609	-14 2.22	10	38.456	-14 13.88	4.14	10.19	5.06	16.17	13.4	20.7	156.3	4.5	121
1987	2	26	.0	2446852.5	10	35.488	-13 55.97	10	37.335	-14 7.61	4.14	10.83	5.07	16.15	13.4	20.7	156.9	4.4	134
1987	2	27	.0	2446853.5	10	34.371	-13 49.62	10	36.218	-14 1.24	4.15	11.48	5.08	16.14	13.4	20.7	157.4	4.3	146
1987	2	28	.0	2446854.5	10	33.260	-13 43.18	10	35.106	-13 54.78	4.16	12.13	5.09	16.12	13.5	20.7	157.8	4.2	154
1987	3	1	.0	2446855.5	10	32.153	-13 36.64	10	33.999	-13 48.22	4.16	12.79	5.10	16.11	13.5	20.7	158.1	4.2	156
1987	3	2	.0	2446856.5	10	31.052	-13 30.02	10	32.897	-13 41.58	4.17	13.44	5.11	16.09	13.5	20.7	158.4	4.1	150
1987	3	3	.0	2446857.5	10	29.957	-13 23.31	10	31.803	-13 34.85	4.18	14.09	5.11	16.08	13.5	20.7	158.6	4.0	140
1987	3	4	.0	2446858.5	10	28.870	-13 16.53	10	30.715	-13 28.05	4.19	14.74	5.12	16.06	13.5	20.8	158.8	4.0	129
1987	3	5	.0	2446859.5	10	27.789	-13 9.67	10	29.634	-13 21.17	4.20	15.39	5.13	16.04	13.5	20.8	158.9	4.0	118
1987	3	6	.0	2446860.5	10	26.717	-13 2.75	10	28.562	-13 14.23	4.20	16.04	5.14	16.03	13.5	20.8	158.9	4.0	106
1987	3	7	.0	2446861.5	10	25.653	-12 55.76	10	27.498	-13 7.21	4.21	16.68	5.15	16.01	13.5	20.8	158.8	4.0	95
1987	3	8	.0	2446862.5	10	24.598	-12 48.71	10	26.443	-13 .14	4.22	17.33	5.16	16.00	13.5	20.8	158.6	4.0	84
1987	3	9	.0	2446863.5	10	23.552	-12 41.60	10	25.397	-12 53.02	4.23	17.97	5.17	15.99	13.6	20.8	158.4	4.1	72
1987	3	10	.0	2446864.5	10	22.516	-12 34.44	10	24.361	-12 45.84	4.24	18.61	5.18	15.97	13.6	20.8	158.1	4.1	61
1987	3	11	.0	2446865.5	10	21.489	-12 27.23	10	23.334	-12 38.61	4.26	19.24	5.19	15.96	13.6	20.8	157.7	4.2	51
1987	3	12	.0	2446866.5	10	20.474	-12 19.98	10	22.319	-12 31.34	4.27	19.87	5.20	15.94	13.6	20.8	157.3	4.2	41
1987	3	13	.0	2446867.5	10	19.469	-12 12.69	10	21.314	-12 24.03	4.28	20.50	5.21	15.93	13.6	20.8	156.8	4.3	31
1987	3	14	.0	2446868.5	10	18.476	-12 5.37	10	20.321	-12 16.68	4.29	21.12	5.22	15.91	13.6	20.8	156.3	4.4	25
1987	3	15	.0	2446869.5	10	17.494	-11 58.01	10	19.339	-12 9.30	4.30	21.73	5.23	15.90	13.6	20.9	155.7	4.5	23

YR	MN	DY	HR	J.D.	R.A. (1950) .0	DEC.	R.A. APPN	DEC.	DELTA	DELDOT	R	RDOT	TMAG	NMAG	THETA	BETA	MOON
1987	3	16	.0	2446870.5	10 16.524	-11 50.63	10 18.369	-12 1.90	4.32	22.35	5.23	15.88	13.6	20.9	155.1	4.6	28
1987	3	17	.0	2446871.5	10 15.567	-11 43.23	10 17.412	-11 54.48	4.33	22.95	5.24	15.87	13.6	20.9	154.4	4.7	37
1987	3	18	.0	2446872.5	10 14.622	-11 35.81	10 16.467	-11 47.04	4.34	23.56	5.25	15.85	13.7	20.9	153.7	4.8	48
1987	3	19	.0	2446873.5	10 13.690	-11 28.38	10 15.535	-11 39.58	4.36	24.15	5.26	15.84	13.7	20.9	152.9	4.9	60
1987	3	20	.0	2446874.5	10 12.771	-11 20.93	10 14.616	-11 32.11	4.37	24.74	5.27	15.82	13.7	20.9	152.2	5.1	73
1987	3	21	.0	2446875.5	10 11.865	-11 13.48	10 13.711	-11 24.64	4.38	25.33	5.28	15.81	13.7	20.9	151.3	5.2	86
1987	3	22	.0	2446876.5	10 10.974	-11 6.02	10 12.819	-11 17.17	4.40	25.91	5.29	15.80	13.7	20.9	150.5	5.3	99
1987	3	23	.0	2446877.5	10 10.096	-10 58.57	10 11.942	-11 9.70	4.41	26.48	5.30	15.78	13.7	20.9	149.7	5.5	112
1987	3	24	.0	2446878.5	10 9.232	-10 51.12	10 11.079	-11 2.23	4.43	27.05	5.31	15.77	13.7	21.0	148.8	5.6	125
1987	3	25	.0	2446879.5	10 8.383	-10 43.69	10 10.230	-10 54.77	4.45	27.61	5.32	15.75	13.8	21.0	147.9	5.7	137
1987	3	26	.0	2446880.5	10 7.549	-10 36.26	10 9.396	-10 47.32	4.46	28.16	5.33	15.74	13.8	21.0	147.0	6.0	155
1987	3	27	.0	2446881.5	10 6.729	-10 28.85	10 8.576	-10 39.89	4.48	28.71	5.34	15.72	13.8	21.0	146.0	6.1	155
1987	3	28	.0	2446882.5	10 5.925	-10 21.46	10 7.772	-10 32.48	4.50	29.25	5.34	15.71	13.8	21.0	145.1	6.3	148
1987	3	29	.0	2446883.5	10 5.135	-10 14.09	10 6.983	-10 25.10	4.51	29.78	5.35	15.70	13.8	21.0	144.1	6.4	138
1987	3	30	.0	2446884.5	10 4.361	-10 6.76	10 6.209	-10 17.74	4.53	30.30	5.36	15.68	13.8	21.0	143.2	6.5	127
1987	3	31	.0	2446885.5	10 3.603	- 9 59.45	10 5.451	-10 10.41	4.55	30.81	5.37	15.67	13.8	21.0	142.2	6.7	115
1987	4	1	.0	2446886.5	10 2.860	- 9 52.18	10 4.709	-10 3.12	4.57	31.31	5.38	15.65	13.8	21.1	141.2	6.8	104
1987	4	2	.0	2446887.5	10 2.134	- 9 44.94	10 3.982	- 9 55.86	4.58	31.80	5.39	15.64	13.9	21.1	140.2	6.9	92
1987	4	3	.0	2446888.5	10 1.423	- 9 37.75	10 3.272	- 9 48.65	4.60	32.29	5.40	15.63	13.9	21.1	139.2	7.1	81
1987	4	4	.0	2446889.5	10 .728	- 9 30.59	10 2.577	- 9 41.48	4.62	32.76	5.41	15.61	13.9	21.1	138.2	7.2	69
1987	4	5	.0	2446890.5	10 .049	- 9 23.49	10 1.898	- 9 34.36	4.64	33.22	5.42	15.60	13.9	21.1	137.2	7.3	59
1987	4	6	.0	2446891.5	9 59.386	- 9 16.44	10 1.236	- 9 27.29	4.66	33.68	5.43	15.59	13.9	21.1	136.2	7.5	48
1987	4	7	.0	2446892.5	9 58.739	- 9 9.43	10 .589	- 9 20.27	4.68	34.12	5.43	15.57	13.9	21.1	135.2	7.6	38
1987	4	8	.0	2446893.5	9 58.108	- 9 2.49	9 59.959	- 9 13.30	4.70	34.56	5.44	15.56	14.0	21.2	134.2	7.7	29
1987	4	9	.0	2446894.5	9 57.494	- 8 55.59	9 59.345	- 9 6.40	4.72	34.98	5.45	15.54	14.0	21.2	133.1	7.8	24
1987	4	10	.0	2446895.5	9 56.895	- 8 48.76	9 58.747	- 8 59.55	4.74	35.39	5.46	15.53	14.0	21.2	132.1	7.9	23
1987	4	11	.0	2446896.5	9 56.313	- 8 41.99	9 58.164	- 8 52.76	4.76	35.80	5.47	15.52	14.0	21.2	131.1	8.0	29
1987	4	12	.0	2446897.5	9 55.746	- 8 35.29	9 57.598	- 8 46.04	4.78	36.19	5.48	15.50	14.0	21.2	130.1	8.2	38
1987	4	13	.0	2446898.5	9 55.196	- 8 28.65	9 57.048	- 8 39.38	4.80	36.57	5.49	15.49	14.0	21.2	129.0	8.3	49
1987	4	14	.0	2446899.5	9 54.661	- 8 22.07	9 56.514	- 8 32.79	4.82	36.94	5.50	15.48	14.0	21.2	128.0	8.4	62
1987	4	15	.0	2446900.5	9 54.143	- 8 15.57	9 55.996	- 8 26.27	4.84	37.31	5.51	15.46	14.1	21.2	127.0	8.5	75
1987	4	16	.0	2446501.5	9 53.640	- 8 9.13	9 55.494	- 8 19.83	4.87	37.66	5.52	15.45	14.1	21.2	126.0	8.6	88
1987	4	17	.0	2446902.5	9 53.153	- 8 2.77	9 55.007	- 8 13.45	4.89	38.01	5.52	15.44	14.1	21.3	125.0	8.7	101
1987	4	18	.0	2446903.5	9 52.682	- 7 56.48	9 54.537	- 8 7.15	4.91	38.34	5.53	15.42	14.1	21.3	123.9	8.8	114
1987	4	19	.0	2446904.5	9 52.227	- 7 50.27	9 54.082	- 8 .92	4.93	38.67	5.54	15.41	14.1	21.3	122.9	8.8	127
1987	4	20	.0	2446905.5	9 51.787	- 7 44.14	9 53.643	- 7 54.78	4.95	38.98	5.55	15.40	14.1	21.3	121.9	8.9	139
1987	4	21	.0	2446906.5	9 51.363	- 7 38.08	9 53.219	- 7 48.70	4.98	39.29	5.56	15.38	14.1	21.3	120.9	9.0	150
1987	4	22	.0	2446907.5	9 50.954	- 7 32.10	9 52.810	- 7 42.71	5.00	39.58	5.57	15.37	14.2	21.3	119.9	9.1	156
1987	4	23	.0	2446908.5	9 50.561	- 7 26.20	9 52.417	- 7 36.80	5.02	39.87	5.58	15.36	14.2	21.3	118.8	9.2	156
1987	4	24	.0	2446909.5	9 50.182	- 7 20.38	9 52.039	- 7 30.97	5.05	40.14	5.59	15.35	14.2	21.4	117.8	9.2	149
1987	4	25	.0	2446910.5	9 49.819	- 7 14.65	9 51.677	- 7 25.22	5.07	40.41	5.60	15.33	14.2	21.4	116.8	9.3	138
1987	4	26	.0	2446911.5	9 49.471	- 7 9.00	9 51.329	- 7 19.56	5.09	40.66	5.60	15.32	14.2	21.4	115.8	9.4	127
1987	4	27	.0	2446912.5	9 49.138	- 7 3.43	9 50.996	- 7 13.99	5.12	40.91	5.61	15.31	14.2	21.4	114.8	9.4	115
1987	4	28	.0	2446913.5	9 48.820	- 6 57.95	9 50.679	- 7 8.50	5.14	41.14	5.62	15.29	14.3	21.4	113.8	9.5	103
1987	4	29	.0	2446914.5	9 48.517	- 6 52.56	9 50.376	- 7 3.09	5.16	41.36	5.63	15.28	14.3	21.4	112.8	9.5	92
1987	4	30	.0	2446915.5	9 48.228	- 6 47.25	9 50.087	- 6 57.78	5.19	41.57	5.64	15.27	14.3	21.4	111.8	9.6	80
1987	5	1	.0	2446916.5	9 47.954	- 6 42.03	9 49.813	- 6 52.55	5.21	41.78	5.65	15.26	14.3	21.4	110.8	9.6	69
1987	5	2	.0	2446917.5	9 47.694	- 6 36.90	9 49.554	- 6 47.41	5.24	41.97	5.66	15.24	14.3	21.5	109.8	9.7	58
1987	5	3	.0	2446918.5	9 47.448	- 6 31.86	9 49.308	- 6 42.37	5.26	42.14	5.67	15.23	14.3	21.5	108.8	9.7	48
1987	5	4	.0	2446919.5	9 47.216	- 6 26.91	9 49.077	- 6 37.41	5.28	42.31	5.67	15.22	14.3	21.5	107.8	9.7	48

3. THE PATH OF COMET HALLEY ON THE SKY, 1985 - 1986

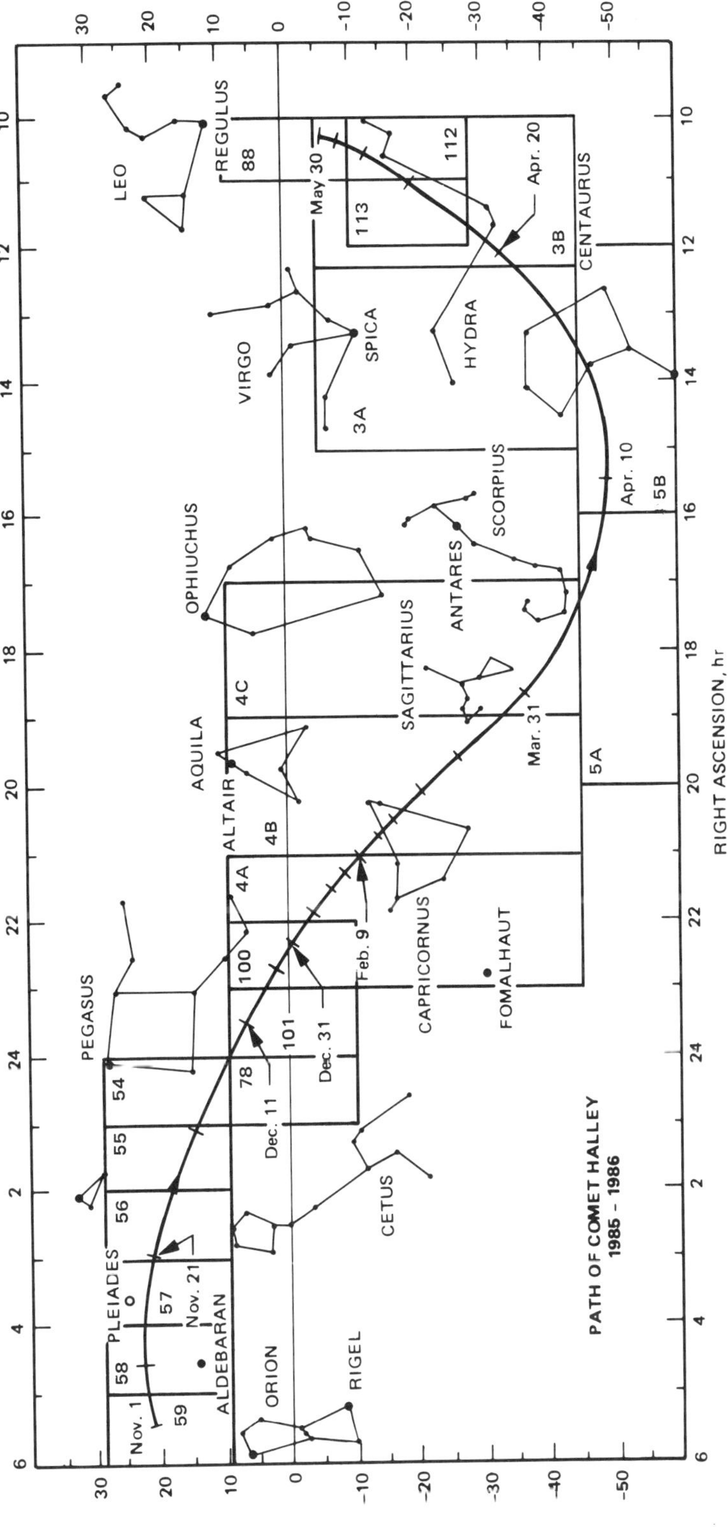

Path of Comet Halley on the Celestial Sphere During November 1985-May 1986. The Charts from the AAVSO and BAA-Tirion Star Atlases Covering the Path are Shown. The Charts Have More Overlap than Indicated

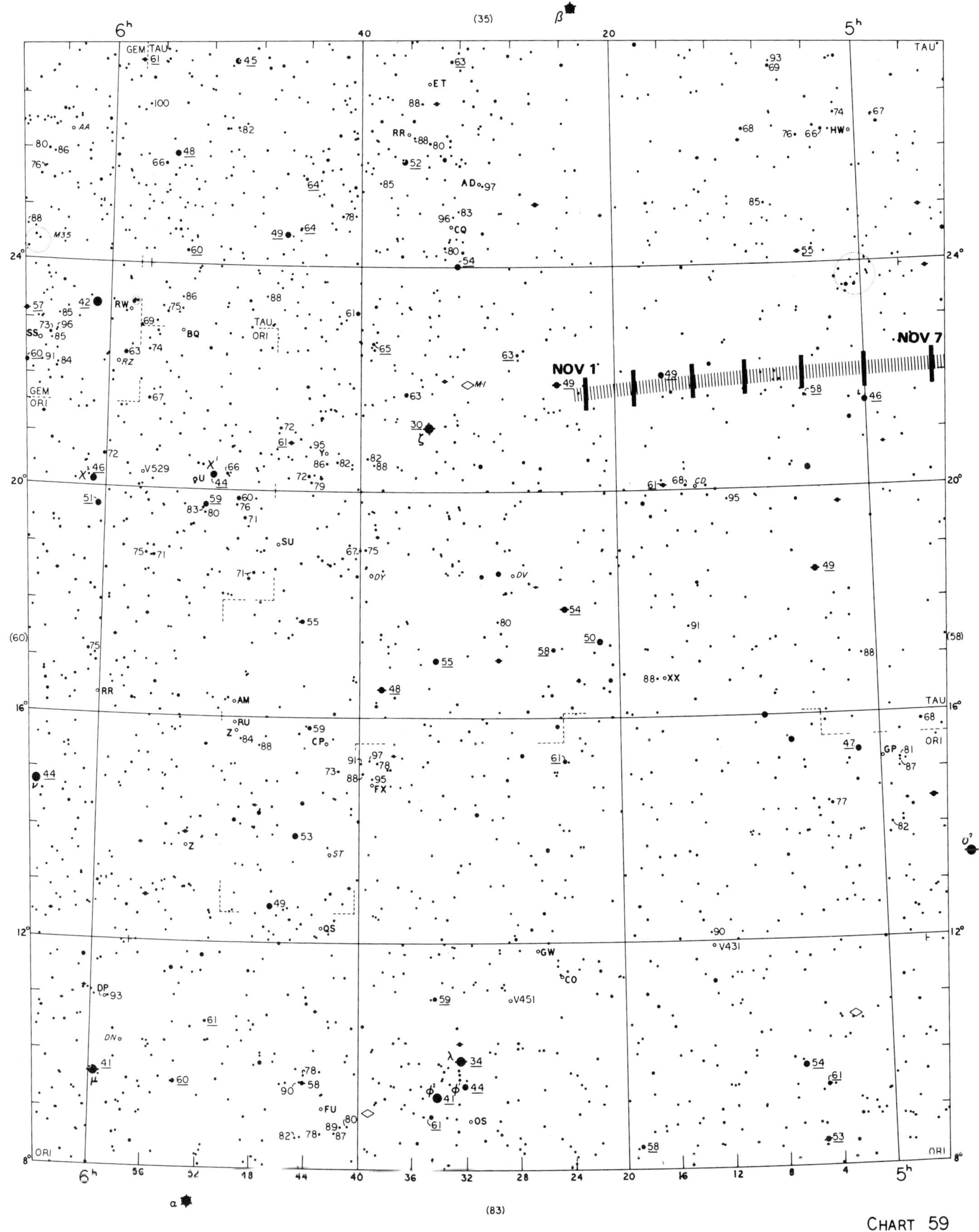

3-5

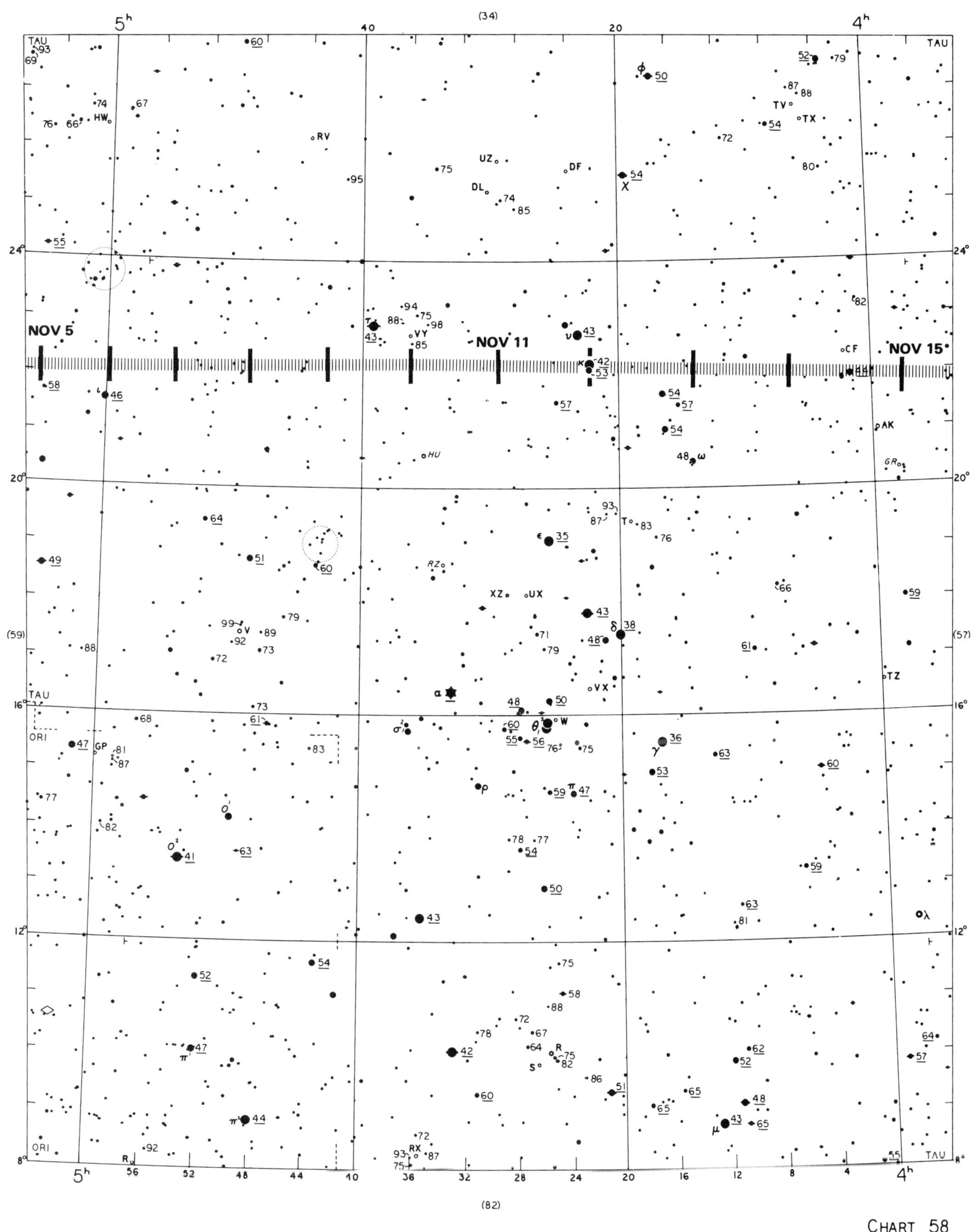

CHART 58

© 1980 AAVSO   REPRODUCED BY SPECIAL PERMISSION OF THE AAVSO

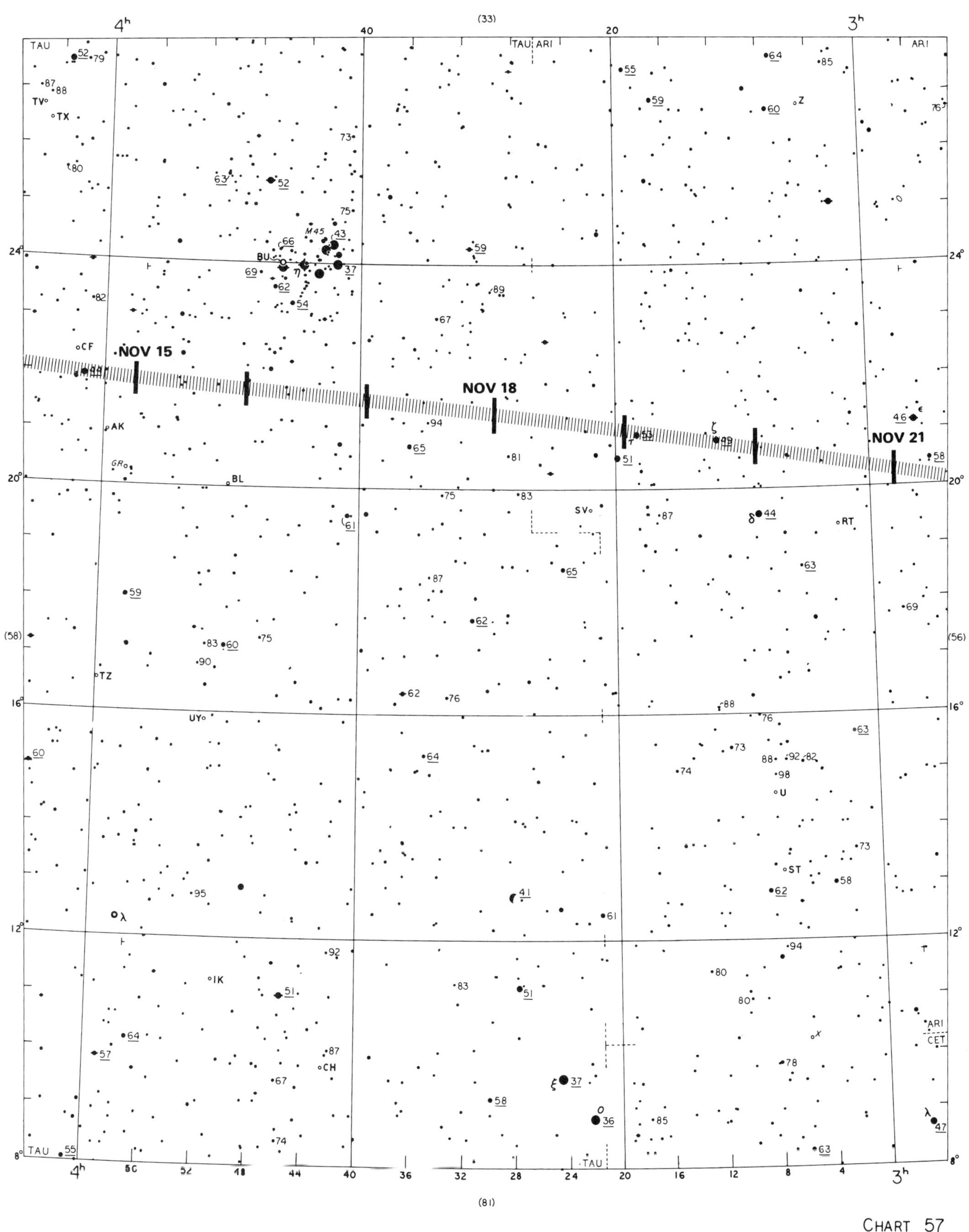

4ʰ
40
(33)
TAU ARI
20
3ʰ
ARI
TAU
52 79
64
85
87 88
55
TV
TX
59
60 Z
76
73
80
63 52
75
M45 43
59
BU
66
89
67
69 η 37
62
54
CF
NOV 15
AK
NOV 18
ζ
46 ε
GR
94
NOV 21
BL
65
81
51
58
75
83
61
SV
87
δ 44
RT
87
65
63
59
62
69
(58)
(56)
83 60 75
90
63
TZ
62 76
88
UY
76
60
73
64
88 92 82
74
98
U
73
95
ST
41
62 58
λ
61
IK
92
94
80
51
83 51
80
64
87
X
ARI
57
CH
78
CET
67
ξ 37
58
O 36
85
λ
74
47
TAU
55
TAU
63
4ʰ
50
52
48
44
40
36
32
28
24
20
16
12
8
4
3ʰ
(81)
CHART 57

3h
40
(32)
20
2h
64
85
ARI
46
RW
Z
X
TRI
36
76
53
62
81
ARI
60
Z
76
70
58
65
50
71
94
82
64
58
71
74
ARI
R 56
82
87
24°
24°
62
78
65
SS
48
65
λ
20
α
92
64
83
κ 50
RX
ν 53
η
ε 46
β
NOV 21
58
0
RT
θ 56
20°
20°
μ
61
63
83
NOV 24
71
89
NOV 26
69
ρ 56
62
51
53 T
65
RW
55
π 91 86
65
88 84
73
64
RY
16°
69
16°
63
68 O
57
TT
σ 55
58
73
ρ
58
90
63
71
92
52
80
S 91
12°
12°
V
93
86
75
84
89
ARI
63
0
ARI
PSC
μ 43
ξ
65
61
λ 47
ξ'
44 ξ
CET
ξ²
56
8° CET
43
8°
3h
56
52
48
44
40
36
32
28
24
20
16
12
8
4
2h
(57)
(80)
(55)
CHART 56

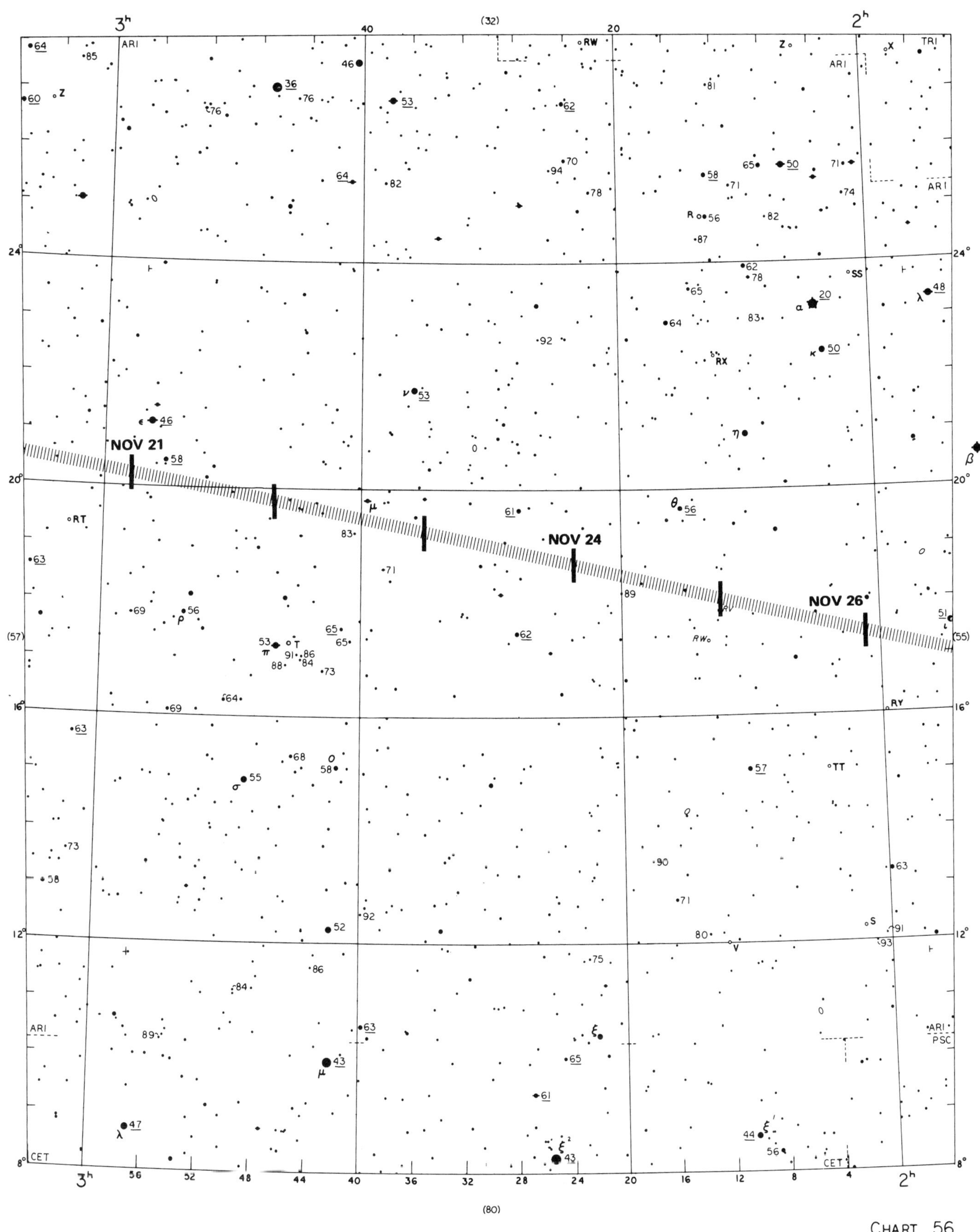

3-11

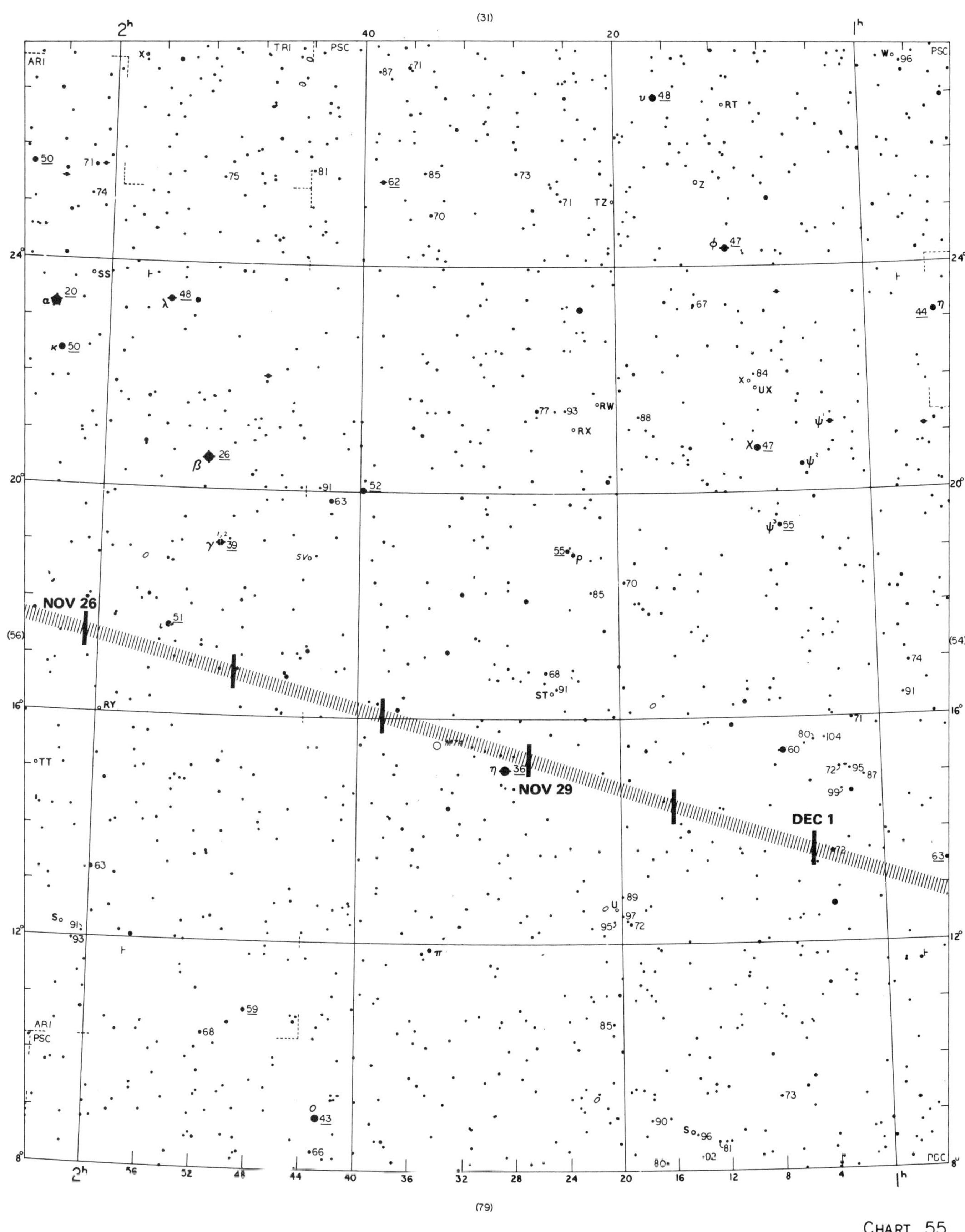

CHART 55

© 1980 AAVSO   REPRODUCED BY SPECIAL PERMISSION OF THE AAVSO

1h
40
(30)
20
0h
PSC AND
AND PEG
W 96
56
88
95
93 84
98 T 81 91
63
58
65
75 77
EE TU
TT
92 DZ 88
89
82 Z
91 95
93
62
46
ψ
24°
24°
ζ
55
44
η GP
AND PEG
ψ'
AND PSC PEG
54
59
ψ¹
20°
20°
χ 48
(55)
(77)
TV
74
91
51
16°
16°
71
61
104
64
0 69 U
98 88 74
72 95 87
58
80
99
γ 28
90 94
T
80
72
63
DEC 2
55
12°
12°
DEC 5
60
PEG PSC
0
63
82 57
DEC 8
8°
PSC
0h
8°
1h
56
52
48
44
40
36
32
28
24
20
16
12
8
4
0h
(78)

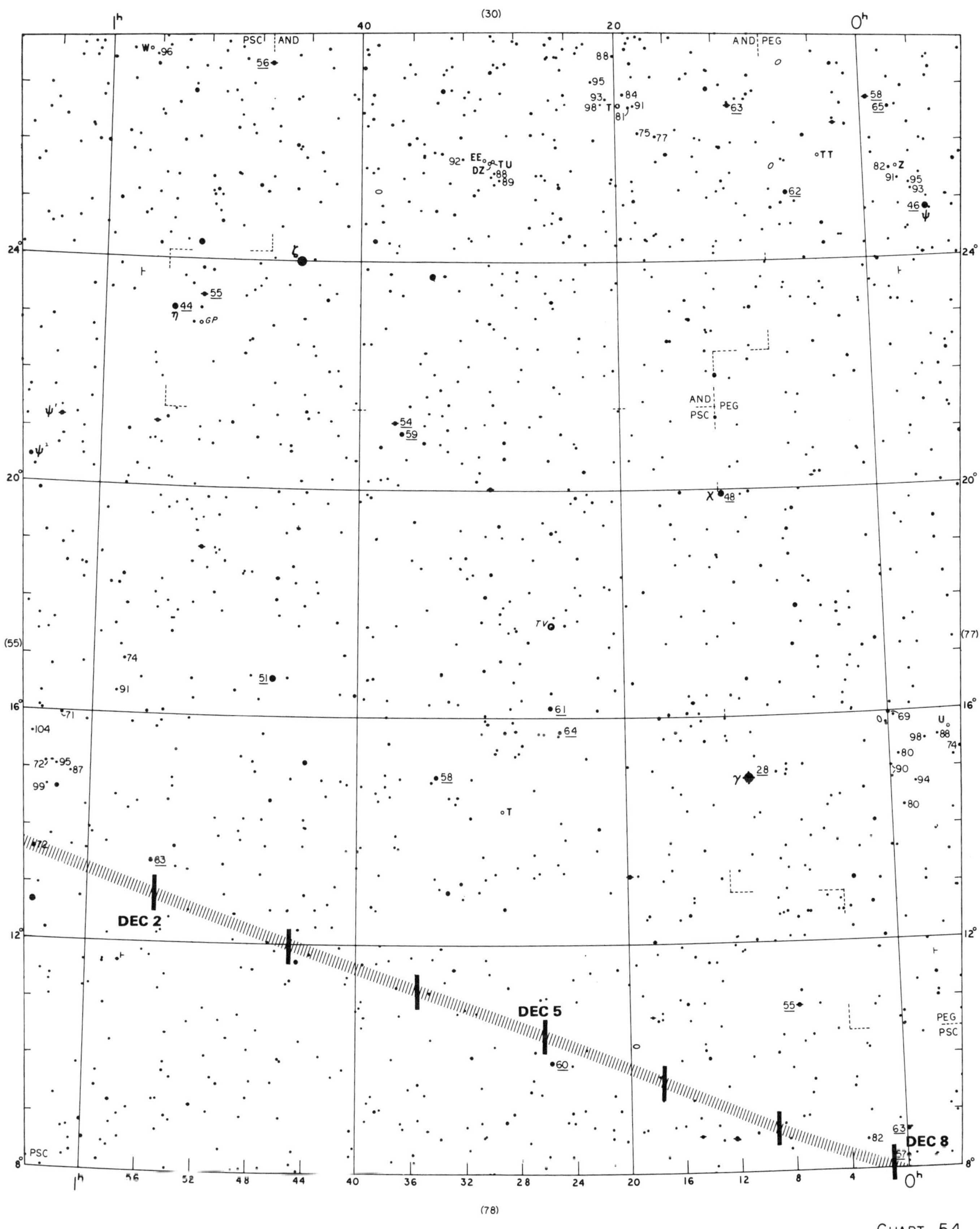

CHART 54

3-15

PSC
1h
40
(54)
20
0h
PSC
10°
10°
DEC 6
DEC 8
82
63
57
54
61
91
57
43
δ 44
59
60
40 ω
6°
6°
60
58
64
PSC
CET
0
2°
2°
61
80
95
70
62
68
0°
0°
0
(79)
67
59
63
(101)
75
48
98
88
0
-2°
-2°
65
70
92
61
61
72
69 64
51
52
91
49
57
59
61
70
58
54
86
46
-6°
-6°
95
44
PSC
0
0
UY
0
51
58
65
0
84
62
35
91
79
60
S 87
85
93
82
-10°
CET
1h
56
52
48
44
40
36
32
28
24
20
16
12
8
4
0h
CET
-10°
(102)
CHART 78

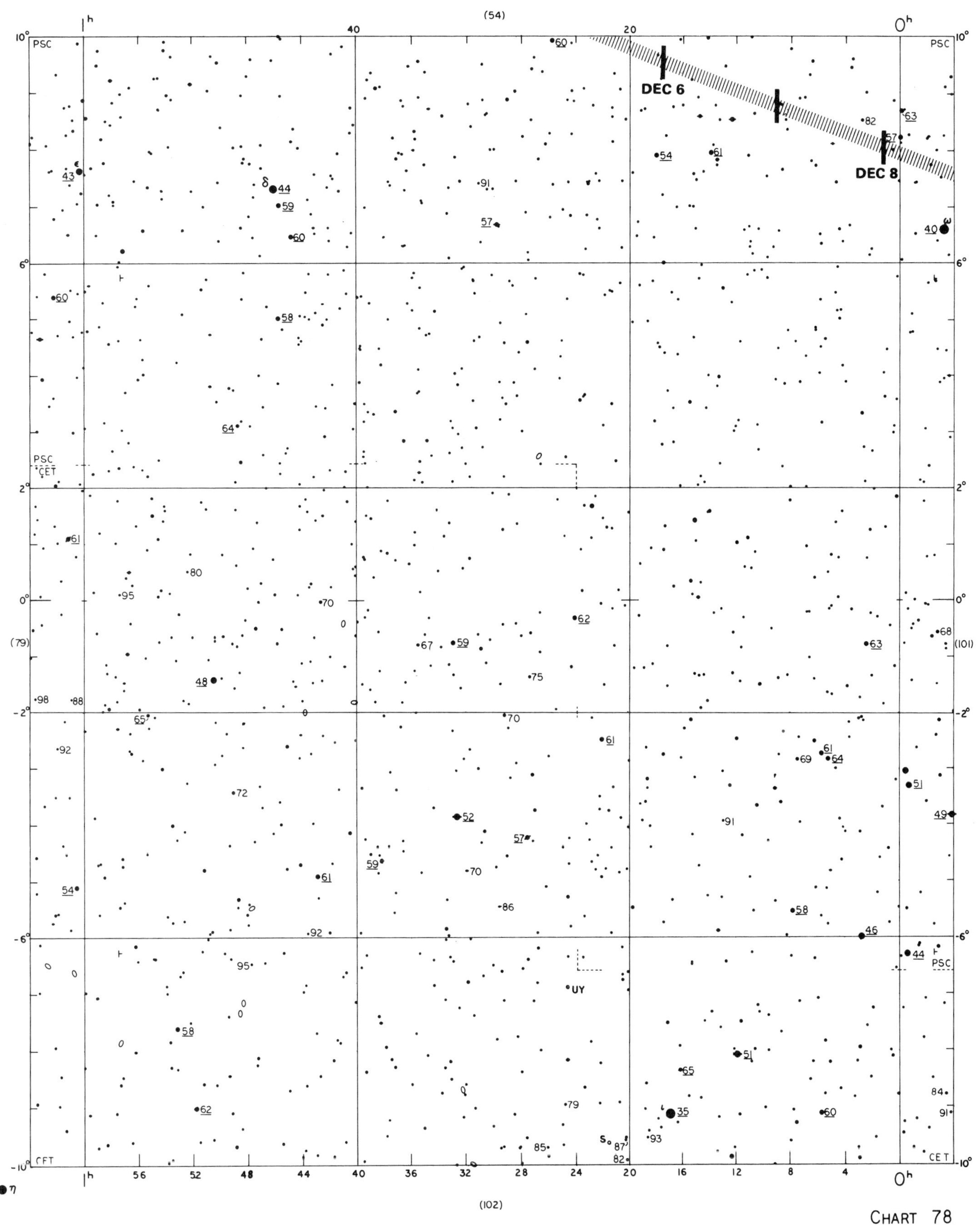

3-17

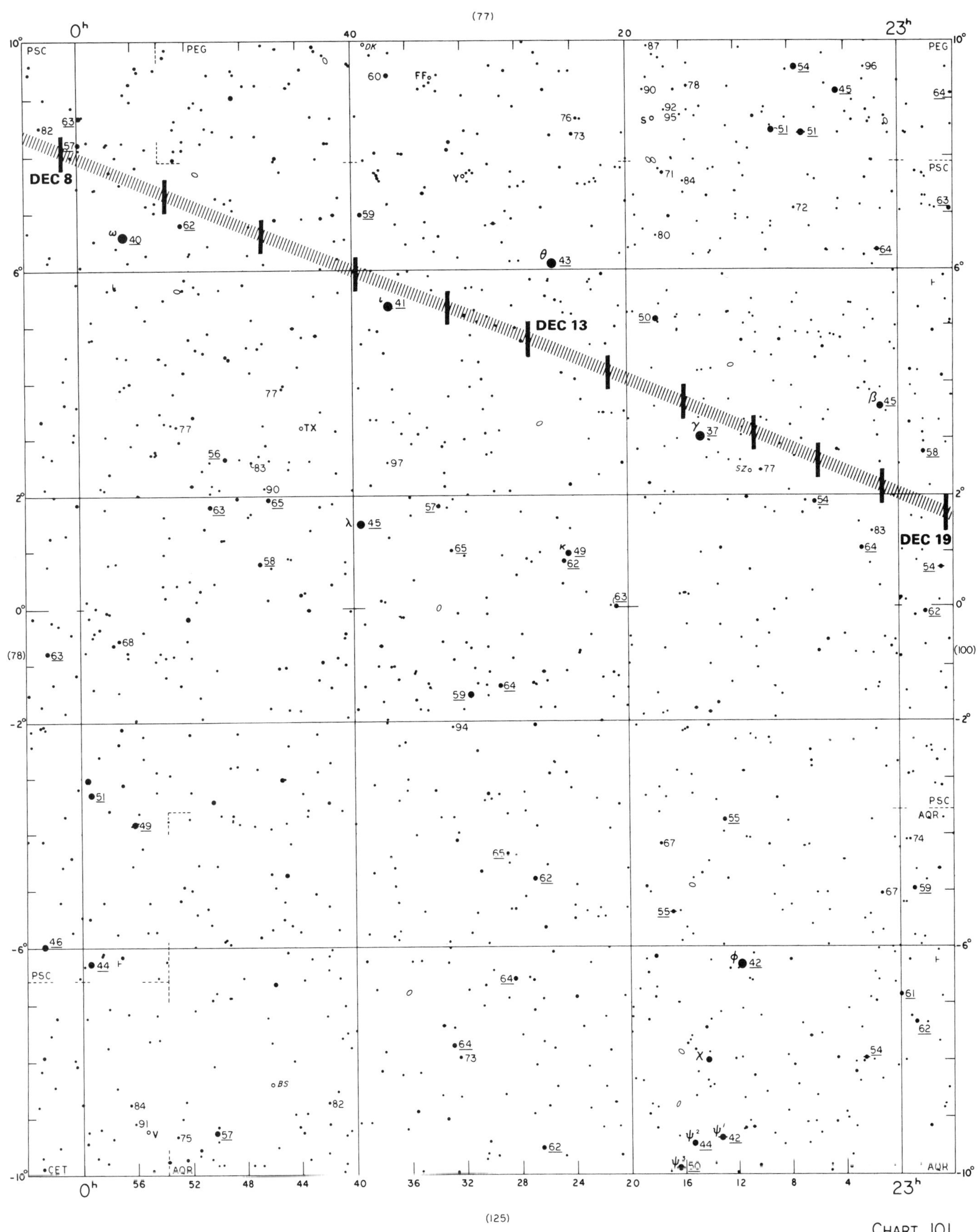

(77)
0h 40 20 23h
10° PSC PEG DEC 8 PEG 10°
DK
60 FF 87
54 96
90 78 45
82 63 92 95 64
57 76 73 S
77 51 51
ω 40 62 Y 71 84 72 PSC
59 80 63
6° θ 43 6°
ι 41 50 DEC 13 64
77
77 TX γ 37 β 45
56 SZ 77 58
83 97
90 65 54
63 57 DEC 19
λ 45 83
65 κ 49 64
58 62 54
63 62
68 0°
(78) 63 (100)
59 64
2° 94 2°
51 PSC
49 AQR
55
67 74
65
62 67 59
55
6° 46 6°
44 φ 42
61
64
62
X 54
64
73
BS
84 82
91 ψ² ψ¹
V 75 57 44 42
CET AQR 62 ψ³ 50 AQR
10° 10°
0h 56 52 48 44 40 36 32 28 24 20 16 12 8 4 23h
(125)
CHART 101

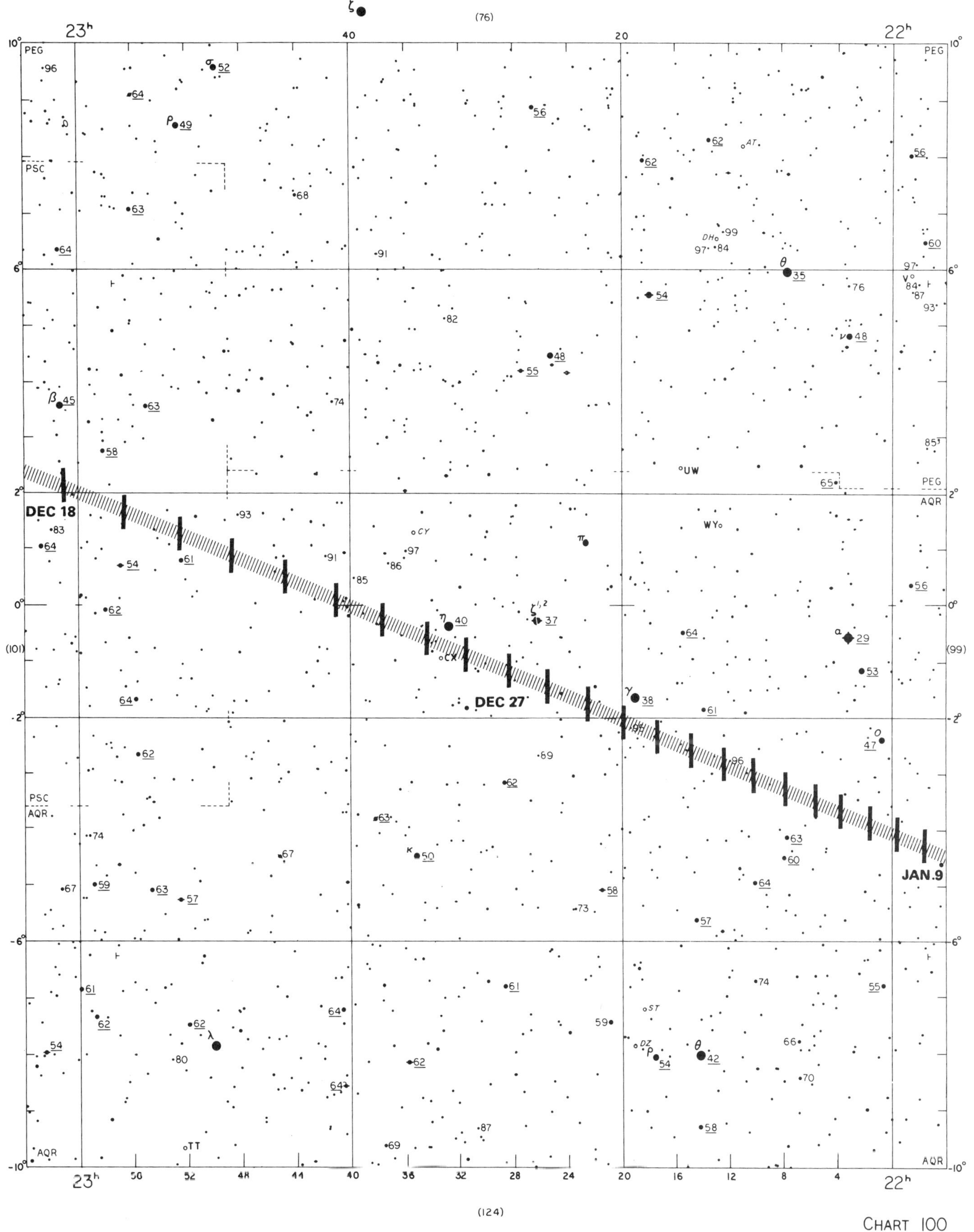

CHART 100

© 1980 AAVSO   REPRODUCED BY SPECIAL PERMISSION OF THE AAVSO

3-21

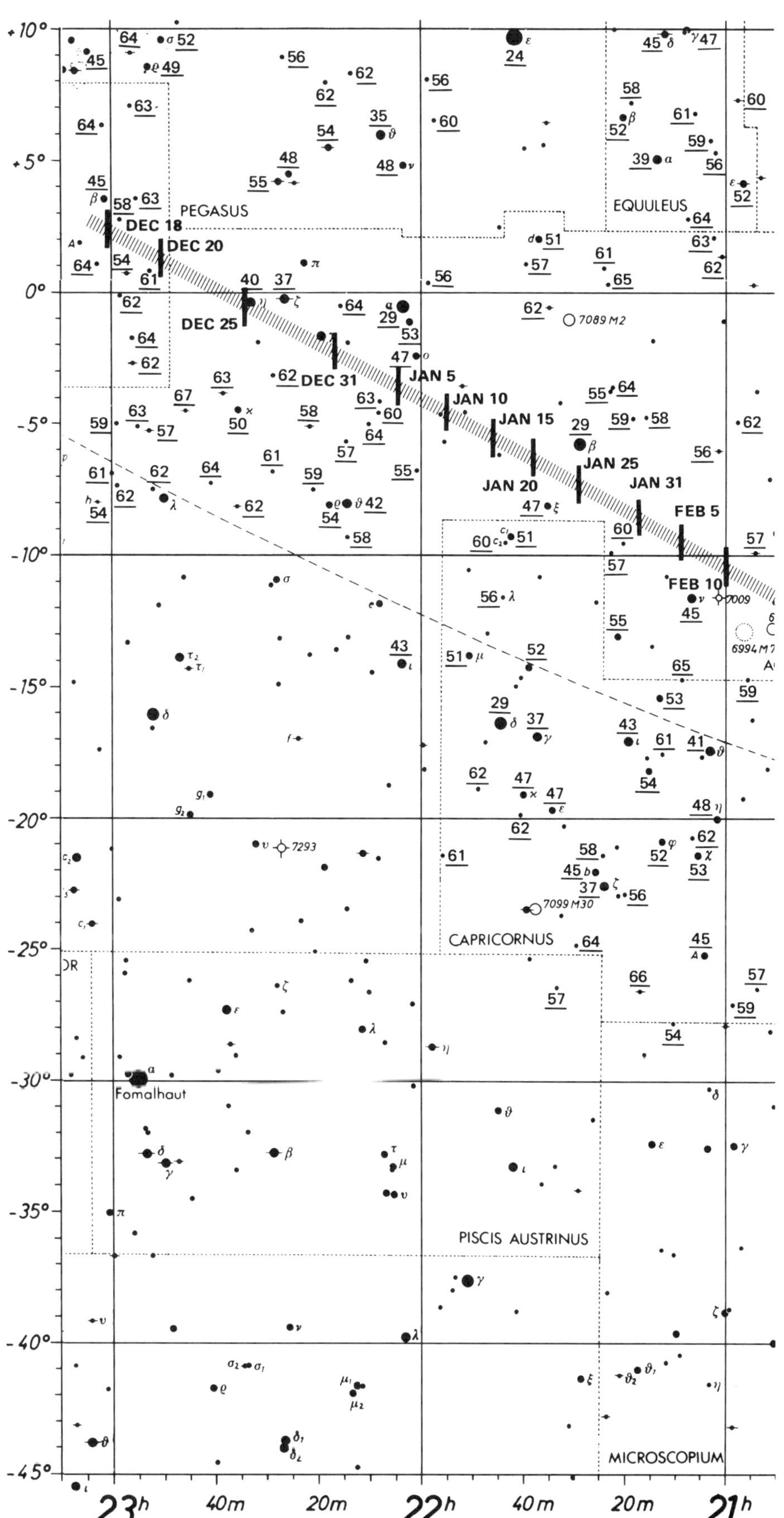

PEGASUS
EQUULEUS
CAPRICORNUS
PISCIS AUSTRINUS
MICROSCOPIUM
Fomalhaut
DEC 18
DEC 20
DEC 25
DEC 31
JAN 5
JAN 10
JAN 15
JAN 20
JAN 25
JAN 31
FEB 5
FEB 10
7089 M2
7099
6994 M?
7293
7099 M30

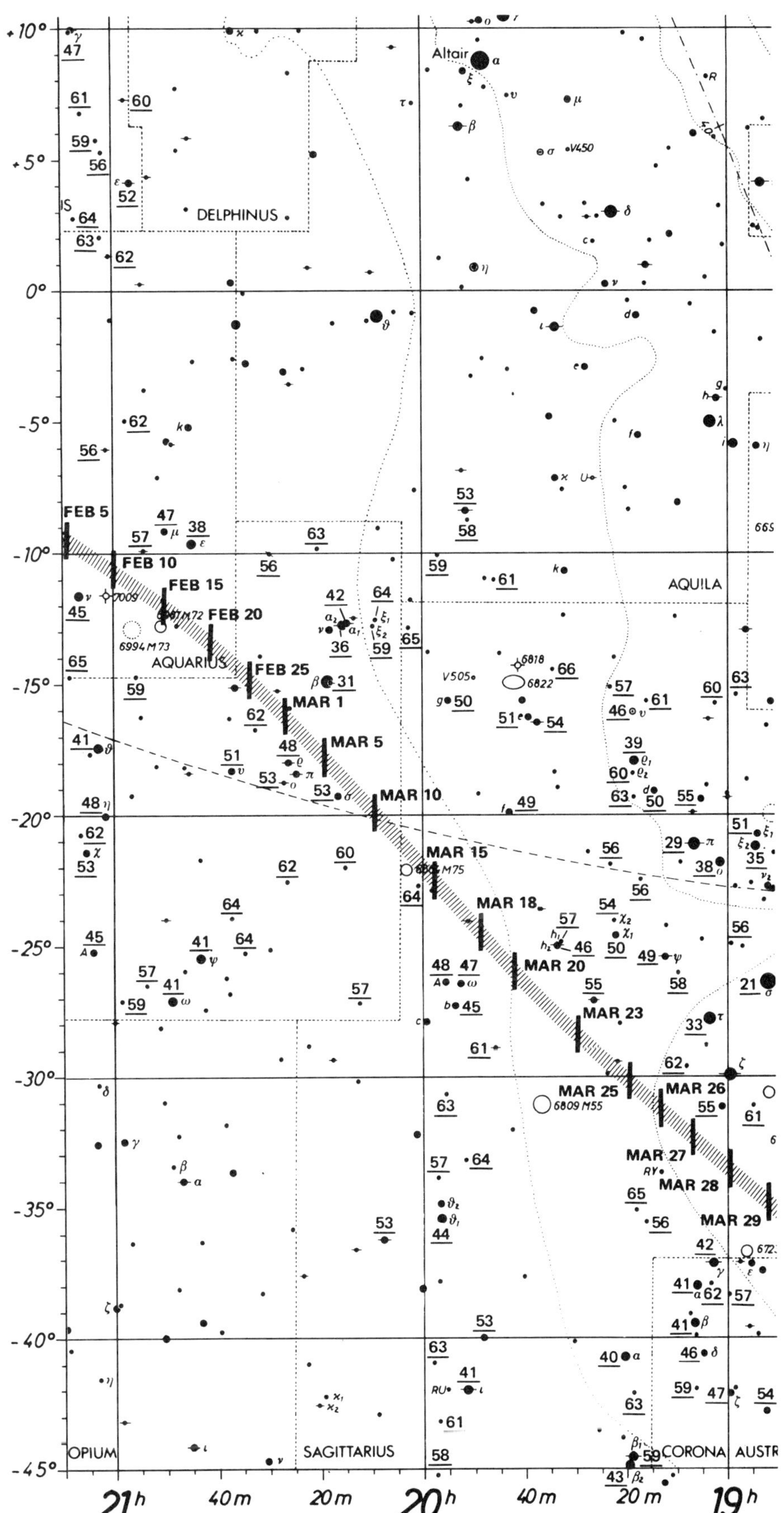

CHART 4B

© 1981 BAA   REPRODUCED BY SPECIAL PERMISSION OF THE BAA

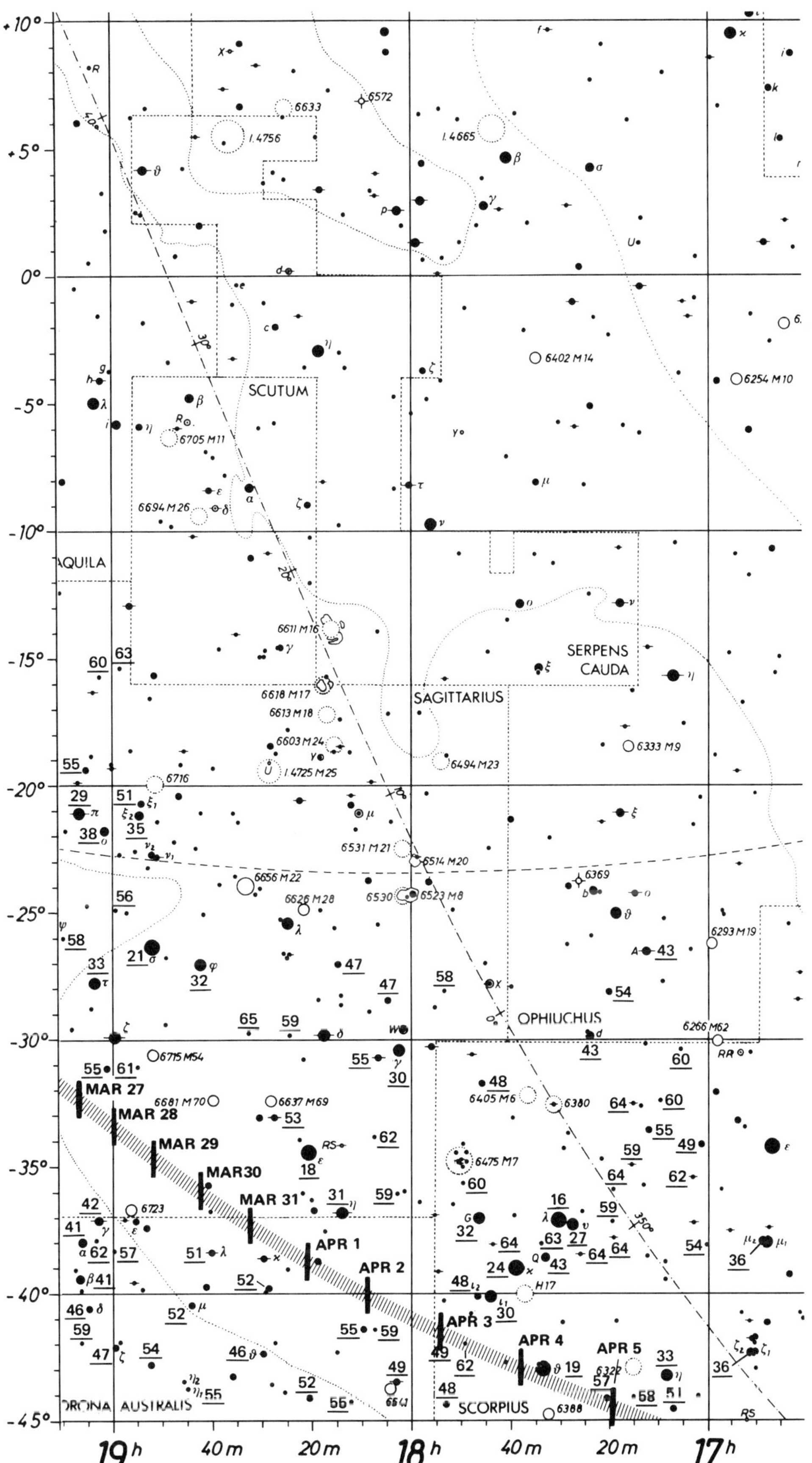

© 1981 BAA   REPRODUCED BY SPECIAL PERMISSION OF THE BAA

20h
53
63
57
64
θ₂
44
θ₁
62
63
RU
53
58
61
41
65
20m
62
57
40
56
63
α
43 β₂ β₁ 39
41
42
46
δ
62
19h
59
41
6723
47
ζ
57
62
54
52
51
65
52 ρ
59
62
55
η₁ η₂
52
μ
λ
40m
6752
58
46
θ
65
50
δ₂
δ₁
APR 1
20m
ω 51
41
35
52
APR 2
46
γ
64
53
55
55
APR 3
61
55
45
6541
CORONA
AUSTRALIS
18h
36 37
θ
49
-40°
61
38
APR 4
48
49
30
48
48
ι
64
H17
40m
46 σ
6385
θ
19
24
16
λ
43 63
υ
TELESCOPIUM
-60°
-50°
53
51
APR 5
64
27
APR 6
58
6322
64
20m
APR 7
63
57
59
APR 8
60
61
APR 9
59
16h
64 34

CHART 5A

3-29

IS
CHAMAELEON
MUSCA
TRIANGULUM AUSTRALE
CIRCINUS
CRUX
NORMA
Acrux
Agena
Toliman
SCO
LUPUS
CENTAURUS
HYDRA
LIBRA
APR 9
APR 10
APR 11
APR 12
APR 13
APR 14
APR 15
-80°
-70°
-60°
-50°
12h   20m   40m   13h   20m   40m   14h   20m   40m   15h
16h
40m
20m

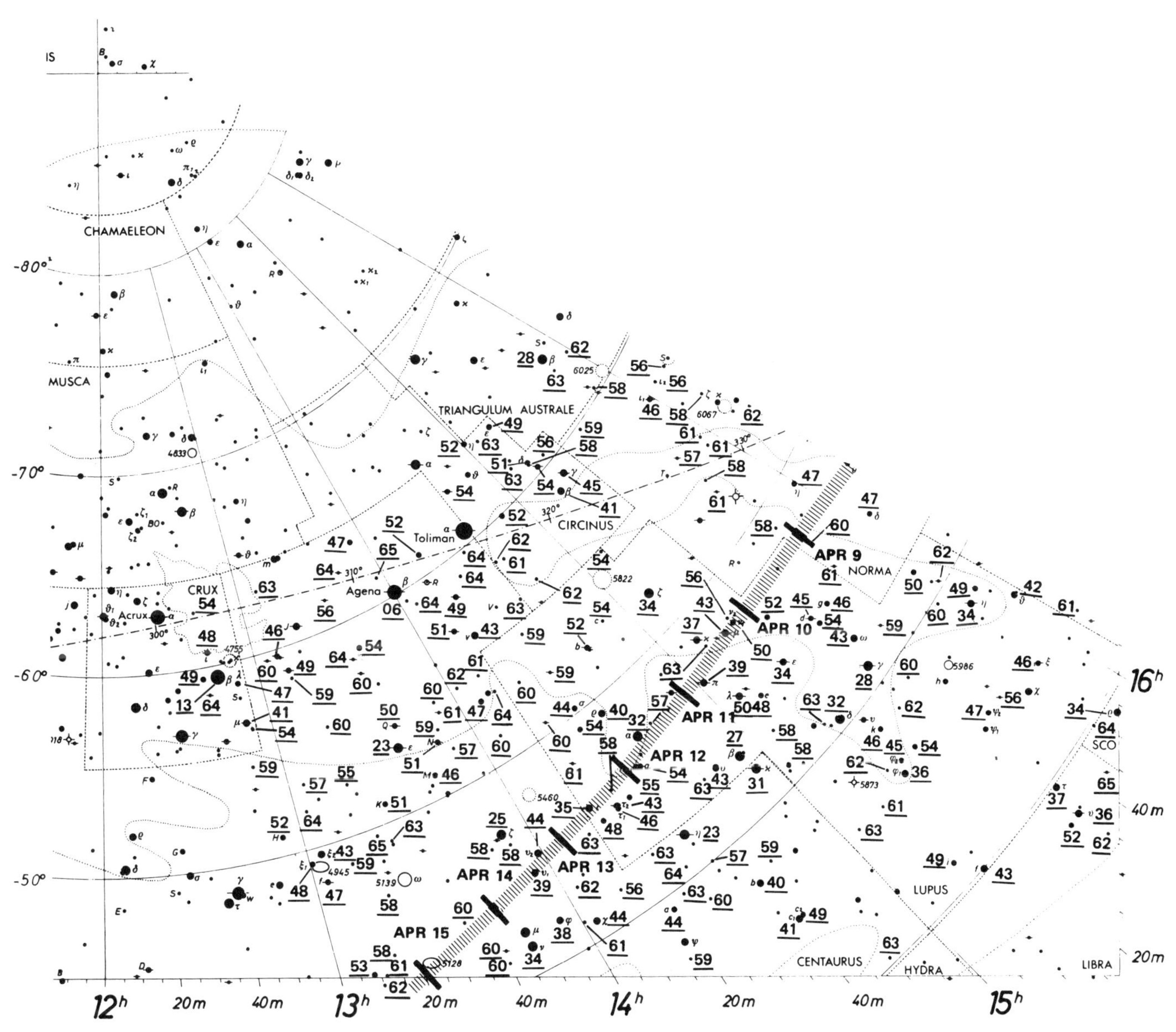

CHART 5B

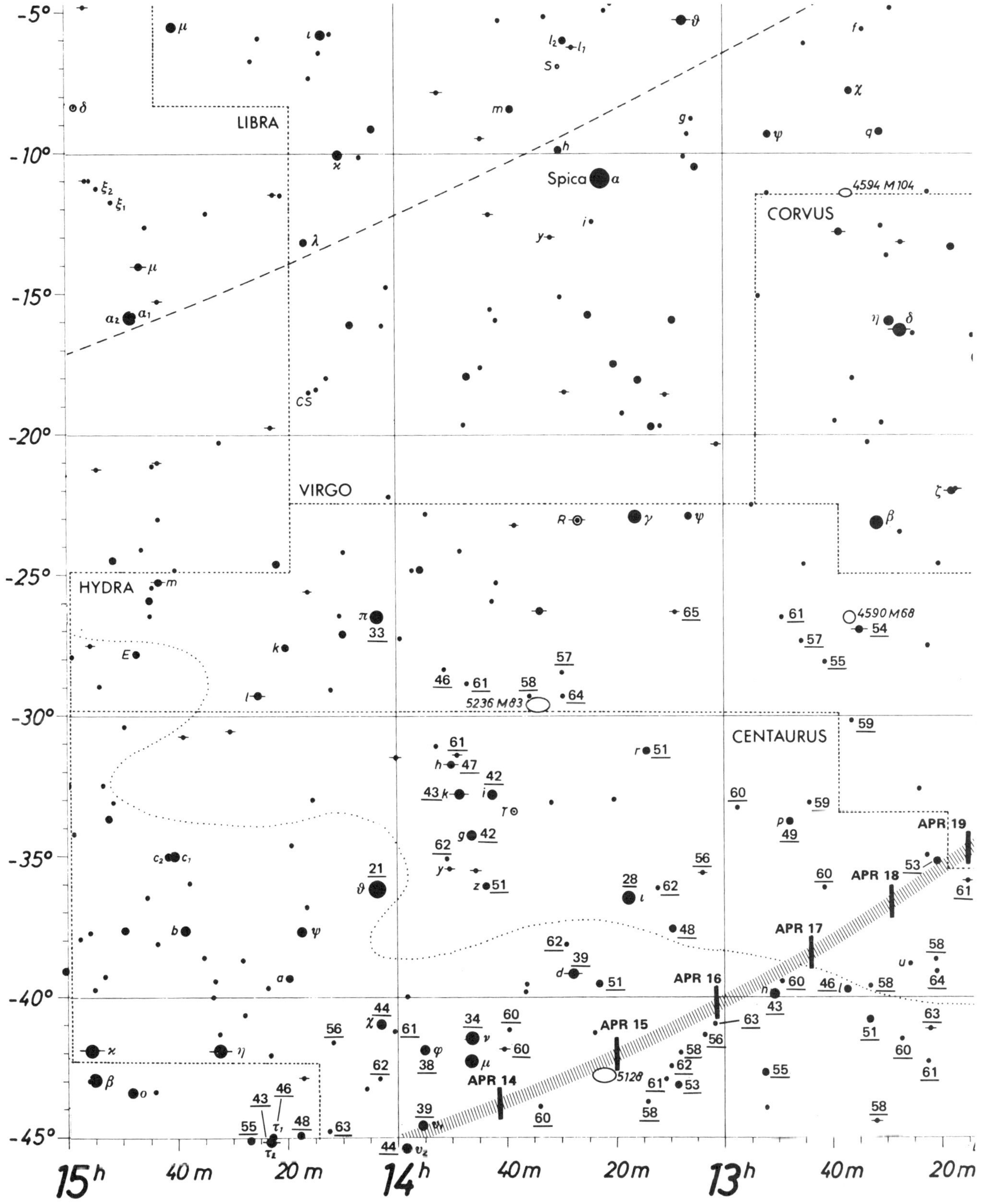

CHART 3A

© 1981 BAA   REPRODUCED BY SPECIAL PERMISSION OF THE BAA

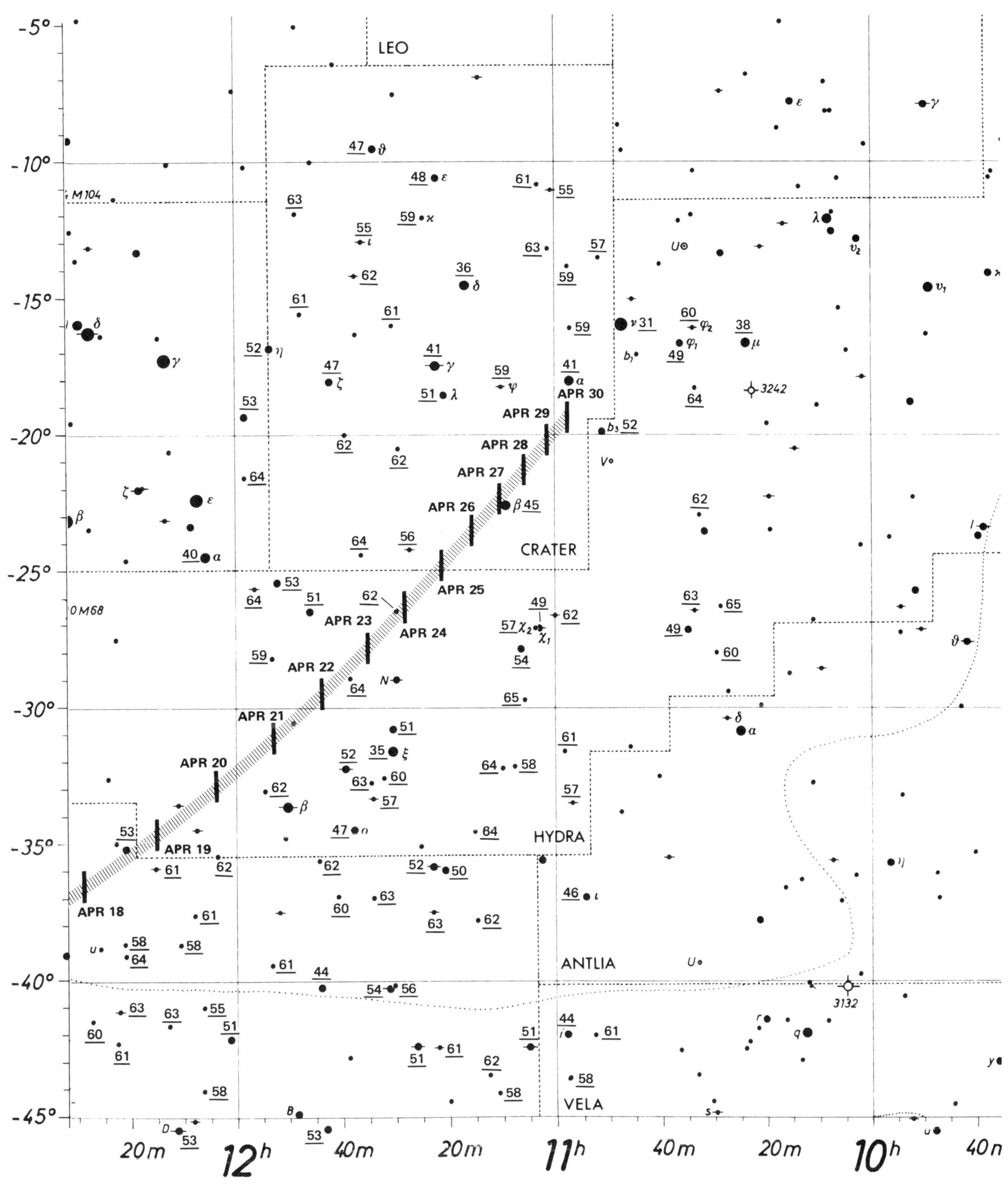
LEO
CRATER
HYDRA
ANTLIA
VELA
M104
M68
APR 30
APR 29
APR 28
APR 27
APR 26
APR 25
APR 24
APR 23
APR 22
APR 21
APR 20
APR 19
APR 18
CHART 3B
20m 12h 40m 20m 11h 40m 20m 10h 40m

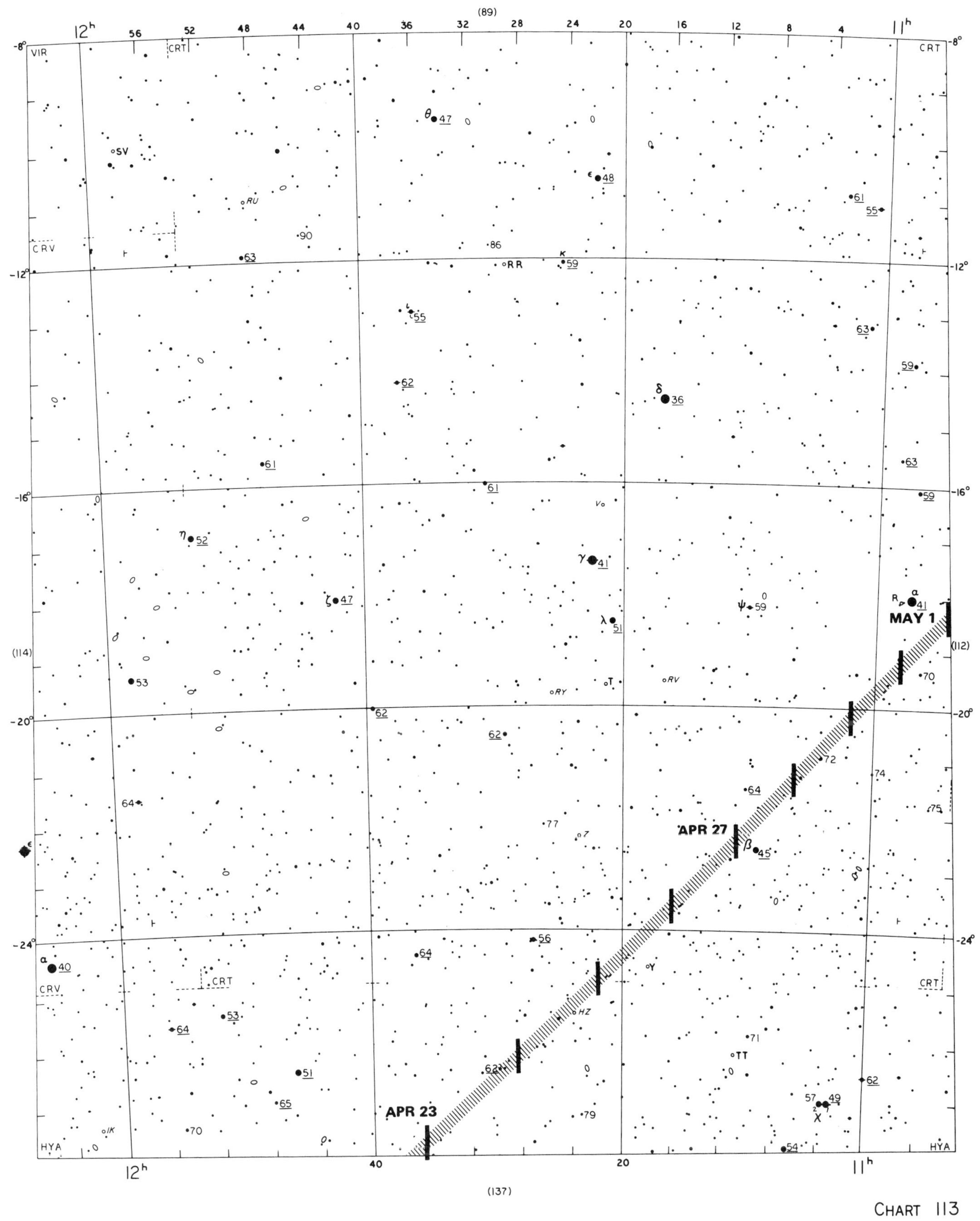
CHART 113
© 1980 AAVSO REPRODUCED BY SPECIAL PERMISSION OF THE AAVSO
APR 23
APR 27
MAY 1
VIR
CRV
HYA
CRT

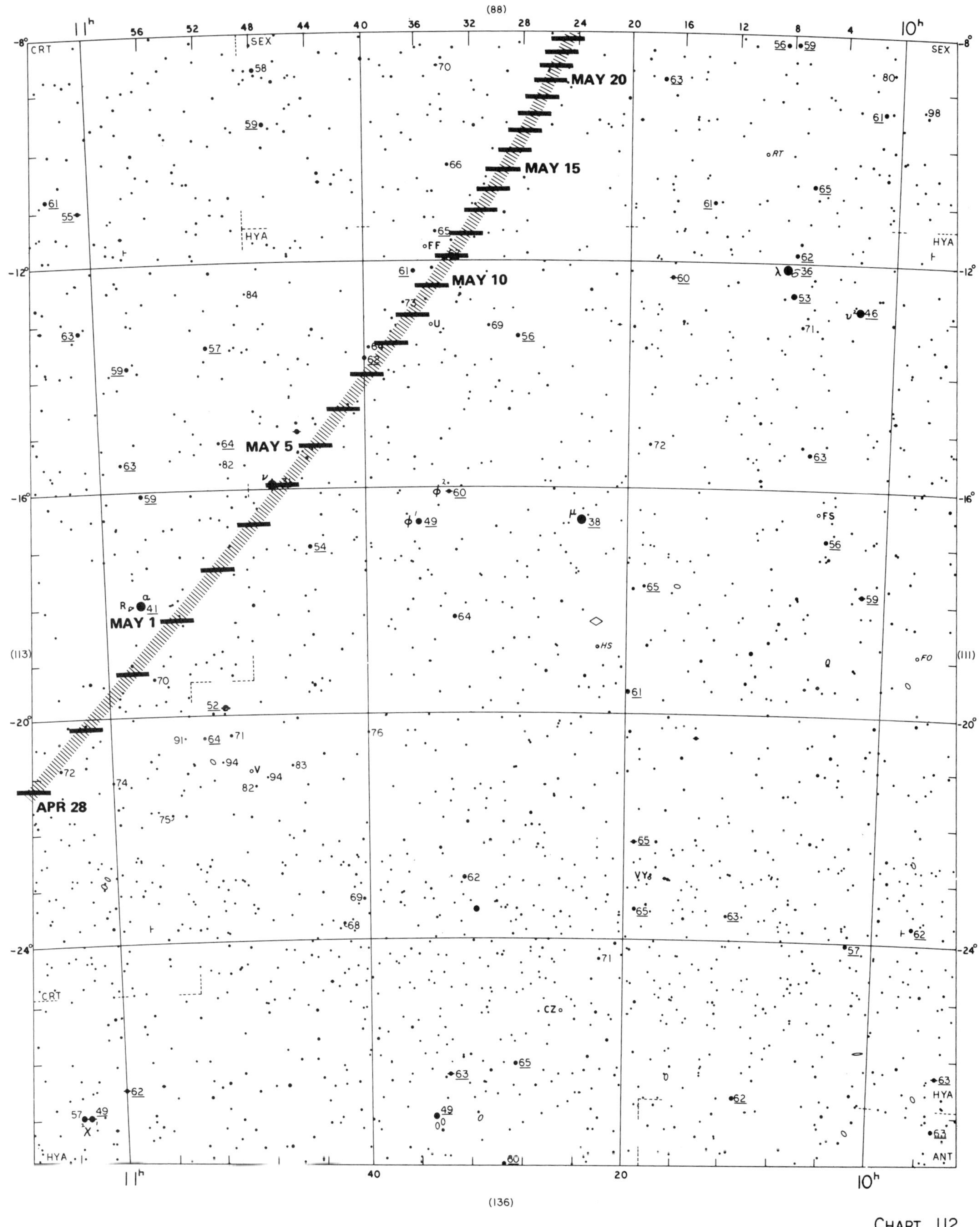

© 1980 AAVSO   REPRODUCED BY SPECIAL PERMISSION OF THE AAVSO

11ʰ
40
(64)
20
10ʰ
10°
LEO
VV
ρ 39
75
97
π 47
χ 46
73
51
85
61
73
77 SEX
50
64
SEX 0
65
62
6°
AP
58
58
48
72
64
70
81
63
70
Z
2°
64
Y
96
99
59
0
95 S
β 51
α 45
63
61
89
89
98
79
94
(89)
(87)
63
55
59
61
δ 52
47
92
SX
60
71
60
93
66
65
76
64
TY
6°
LEO
79
98
MAY 30   X
84
62
72   62
86
MAY 25
62
87
ε 52   70
56   59
70
MAY 20
80
58
63
59
MAY 17
61
98
CRT
SEX
SEX
11ʰ 56 52 48 44
40 36 32 28 24
20 16 12 8 4
10ʰ
(112)
CHART 88

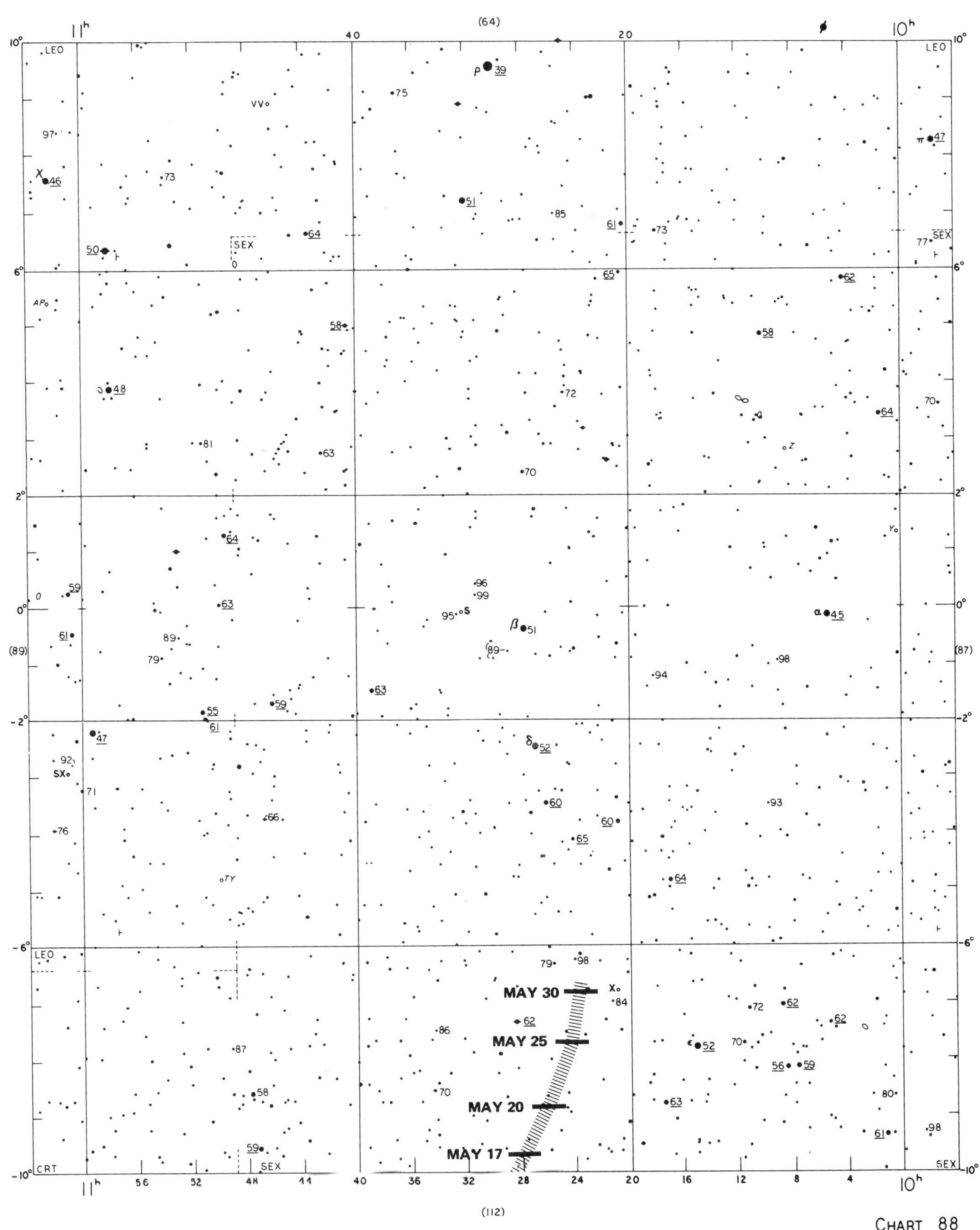

3-41

4. CALIBRATION AND STANDARD STARS

STANDARD STARS FOR IHW PHOTOMETRY

TABLE 4-1

Primary Equatorial Flux Standards

Star Identification

HD	Other	R. A. 1950	Dec.	V Mag.	B-V	Spectrum
		h m s	° ′ ″			
3379	53 Psc	0 23 47.7	+ 14 57 24	5.88	-0.15	B2.5IV
26912	μ Tau	4 12 49.0	+ 8 46 07	4.27	-0.07	B3IV
52266	BD - 5° 1912	6 57 53.9	- 5 45 21	7.23	-0.01	O9V
74280	η Hya	8 40 36.7	- 3 34 46	4.29	-0.20	B4V
89688	(RS) 23 Sex*	10 18 27.1	+ 2 32 31	6.68	-0.09	B2.5IV
120086	BD -1° 2858	13 44 44.2	- 2 11 40	7.89	-0.18	B3III
120315	η UMa	13 45 34.3	+ 49 33 44	1.86	-0.19	B3V
149363	BD -5° 4318	16 31 47.9	- 6 01 59	7.80	+0.01	B0.5III
164852	96 Her	18 00 14.7	+ 20 49 56	5.27	-0.09	BV3
191263	BD +10° 4189	20 06 15.1	+ 10 34 44	6.33	-0.14	B3IV
219188	BD +4° 4985	23 11 28.0	+ 4 43 29	6.9	-	B0.5III

* Although it has a variable star designation, several more recent
  investigations find $\Delta m < 0\overset{m}{.}01$.

TABLE 4-2A

Primary Solar Analogs

HD	Other	R. A. 1950	Dec.	V Mag.	B-V	Spectrum[+]
		h m s	° ′ ″			
28099	Hyades - vB 64	4 23 47.7	+16 38 07	8.12	+0.66	G6V
29461	Hyades - vB 106	4 36 07.6	+14 00 29	7.96	+0.66	G5V
30246	Hyades - vB 142	4 43 38.9	+15 22 59	8.33	+0.67	G5V
44594	HR (YBS) 2290	6 18 47.1	-48 42 50	6.60	+0.66	G2V
105590*	BD -11° 3246	12 06 53.2	-11 34 36	6.56	+0.66	G2V
186427**	16 Cyg B	19 40 32.0	+50 24 03	6.20	+0.66	G5V

+ Spectral types are from various sources and apparently indicate diferences
  in the classifiers rather than in the stars.

* Brightest member of a triple system; $8\overset{m}{.}9$ and $9\overset{m}{.}1$ companions at
  (1$\overset{s}{.}$6 E, 4″ N) and (0$\overset{s}{.}$5 W, 22′ S).

** Fainter member of a double system, 6^{m} companion at (3^{S} W, 28″ N).

From Hardorp (1982 A&A <u>105</u>, 120 and references therein).

TABLE 4-2B

Other Solar Analogs

Star Identification

HD	Other	R.A. 1981		Dec.		V. Mag.
		h m		° '		
--	SAO 120107	13 46.0		+ 6	07	9.26
120528	SAO 28894	13 48.0		+53	20	8.55
144873	SAO 65083	16 06.0		+34	09	8.84
--	+15° 3364	18 06.5		+15	57	8.64
191854	SAO 49262	20 09.6		+43	53	7.43

TABLE 4-3

Faint Equatorial Solar Analogs

Landolt #[*]	V Mag.	Spectrum	R. A. 1975		Dec.	
			h m		° '	
92-433	11.65	G2	0 55.6		+0	53
93-241	9.39	G2	1 54.0		+0	29
94-293	7.02	G5	2 54.0		+0	20
95-236	11.48	G2	3 54.9		+0	04
96-393	9.66	G1	4 51.2		-0	00
97-249	11.74	G2	5 55.8		+0	01
98-313	11.07	G0	6 51.5		-0	34
99-358	9.59	G3	7 52.7		-0	18
100-289	9.13	G7	8 52.5		-0	27
101-057	11.58	G3	9 56.8		-0	54
102-1081	9.91	gG2	10 55.8		-0	05
103-487	11.84	G3	11 53.9		-0	15
104-335	11.70	G3	12 41.1		-0	25
105-257	9.14	G0	13 37.1		-0	52
106-1146	9.10	dG2	14 42.5		+0	09
107-469	12.17	G0	15 38.4		-0	22
108-1911	8.04	G3	16 36.5		+0	06
109-381	11.71	G0	17 42.9		-0	20
110-529	11.41	G0	18 42.7		+0	25
111-1342	9.22	gK2	19 36.0		+0	13
112-636	9.85	G3	20 40.3		+0	11
113-459	12.13	G0	21 40.0		+0	36
114-651	10.28	G2	22 39.9		+0	55
115-366	12.11	dG3	23 43.0		+0	53

* c.f. A.U. Landolt, 1973, AJ 78, 959.

TABLE 4-4

Suggested Spectroscopic Calibration Stars

Star	V Mag.	R.A. (h m)		Dec. (° ′)		Name
β Ari	2.65	1	53.6	+20	43	Sheratan
β Tri	3.00	2	08.4	+34	54	
γ Cet AB	3.48	2	42.2	+03	10	
β Eri	2.79	5	06.9	−05	06	
β Aur	1.86	5	58.0	+44	57	Menkalinan
α CMa A	−1.47	6	44.2	−16	42	Sirius
δ Vel AB	1.95	8	44.2	−54	38	
ι UMa A	3.12	8	57.9	+48	07	
β Car	1.67	9	13.0	−69	38	Miaplacidus
β UMa	2.37	11	00.6	+56	30	Merak
β Leo	2.14	11	48.0	+14	41	Denebola
γ Cen AB	2.17	12	40.5	−48	51	
ζ Vir	3.37	13	33.7	−00	30	
α Lib A	2.76	14	49.8	−15	54	Zubenelgenubi
γ Tr A	2.89	15	17.1	−68	36	
α Cr B	2.23v	15	33.8	+26	47	Alphecca
η Oph AB	2.43	17	09.3	−15	42	Sabik
δ Her	3.14	17	14.2	+24	51	
α Lyr	0.04	18	36.2	+38	46	Vega
ζ Sgr AB	2.61	19	01.3	−29	54	
ζ Aql A	2.99	19	04.5	+13	50	
α Aql	0.77	19	49.8	+ 8	49	Altair
β Pav	3.45	20	43.2	−66	17	
δ Cap	2.92v	21	45.9	−16	13	
α PsA	1.15	22	56.5	−29	44	Fomalhaut

Adapted from the RASC Observer's Handbook 1982

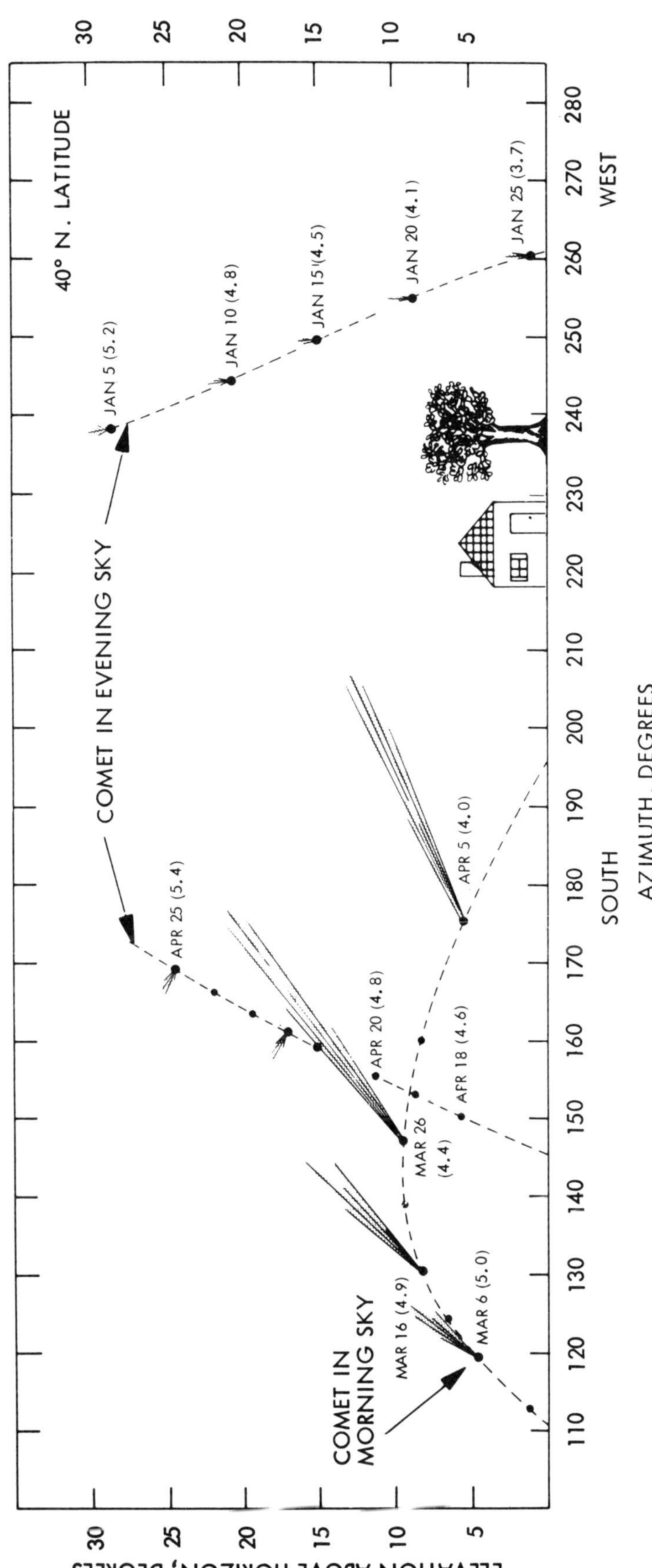

Comet Halley Observing Conditions in 1986 for Observers Located at 40° North Latitude. Comet Positions are Given for Beginning of Morning Astronomical Twilight or End of Evening Astronomical Twilight. Approximate Total Visual Magnitudes are Given in Parentheses Following Dates. Viewing with Binoculars and Ideal Observing Conditions are Assumed.

According to a recent re-examination of past apparitions, the comet may be as much as two magnitudes brighter than some of the figures given here. See the January 1984 issue of Sky & Telescope for details. Reprinted from The Comet Halley Handbook, by Donald Yeomans.

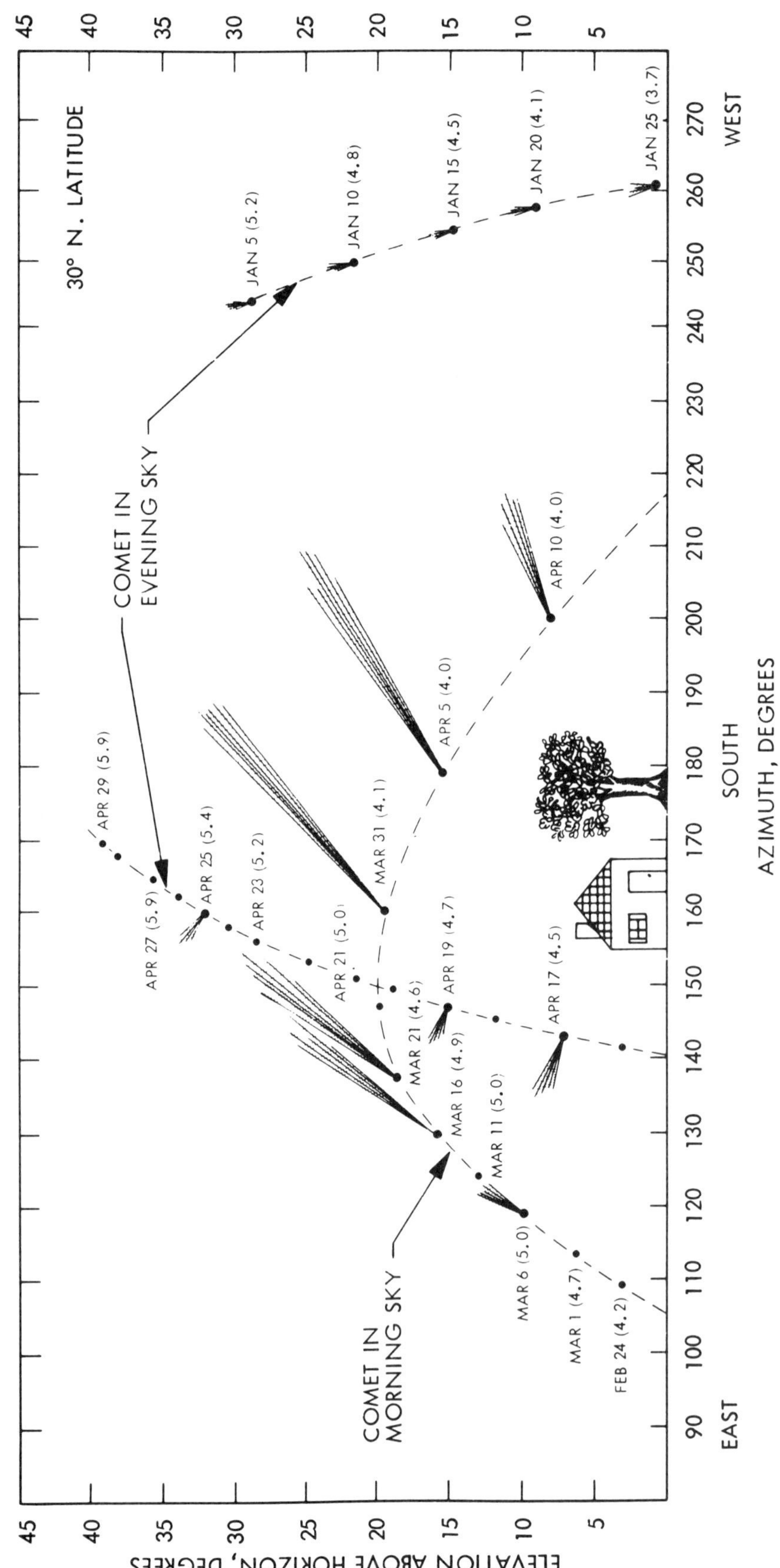

Comet Halley Observing Conditions in 1986 for Observers Located at 30° North Latitude. Comet Positions are Given for Beginning of Morning Astronomical Twilight or End of Evening Astronomical Twilight. Approximate Total Visual Magnitudes are Given in Parentheses Following Dates. Viewing with Binoculars and Ideal Observing Conditions are Assumed.

According to a recent re-examination of past apparitions, the comet may be as much as two magnitudes brighter than some of the figures given here. See the January 1984 issue of Sky & Telescope for details. Reprinted from The Comet Halley Handbook, by Donald Yeomans.

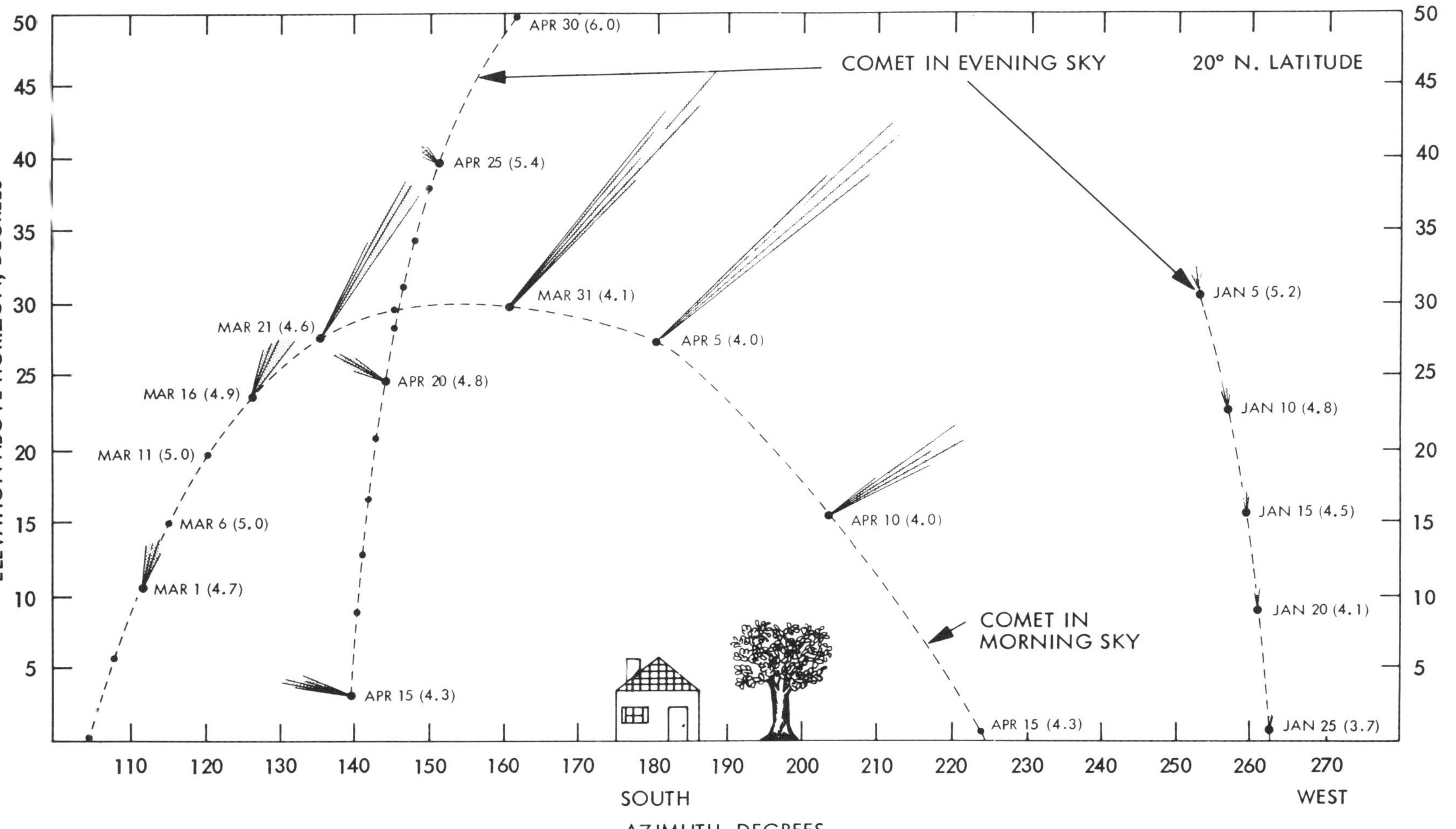

Comet Halley Observing Conditions in 1986 for Observers Located at 20° North Latitude. Comet Positions are Given for Beginning of Morning Astronomical Twilight or End of Evening Astronomical Twilight. Approximate Total Visual Magnitudes are Given in Parentheses Following Dates. Viewing with Binoculars and Ideal Observing Conditions are Assumed.

According to a recent re-examination of past apparitions, the comet may be as much as two magnitudes brighter than some of the figures given here. See the January 1984 issue of Sky & Telescope for details.
Reprinted from The Comet Halley Handbook, by Donald Yeomans.

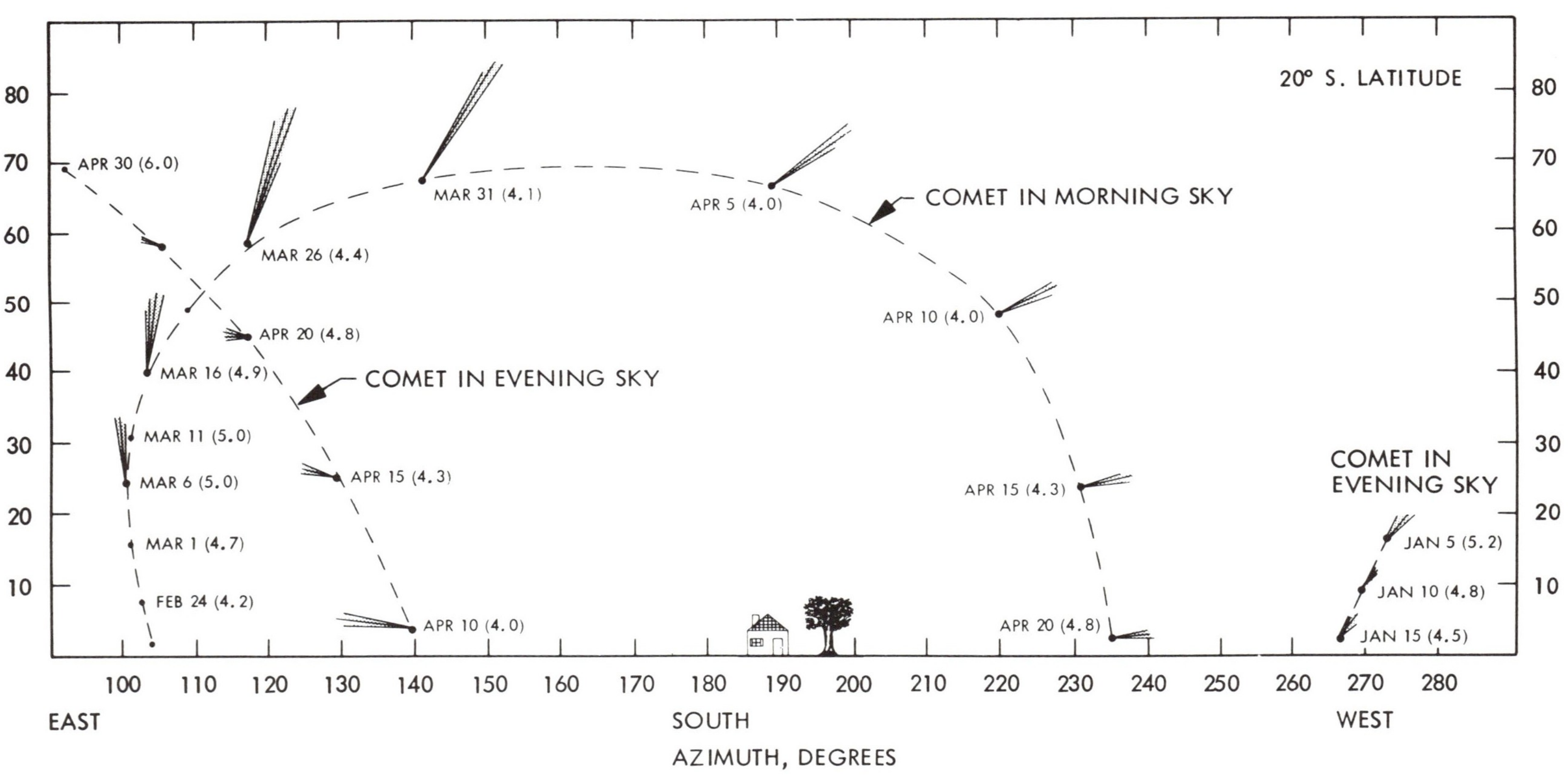

Comet Halley Observing Conditions in 1986 for Observers Located at 20° South Latitude. Comet Positions are Given for Beginning of Morning Astronomical Twilight or End of Evening Astronomical Twilight. Approximate Total Visual Magnitudes are Given in Parentheses Following Dates. Viewing with Binoculars and Ideal Observing Conditions are Assumed.

According to a recent re-examination of past apparitions, the comet may be as much as two magnitudes brighter than some of the figures given here. See the January 1984 issue of Sky & Telescope for details.
Reprinted from The Comet Halley Handbook, by Donald Yeomans.

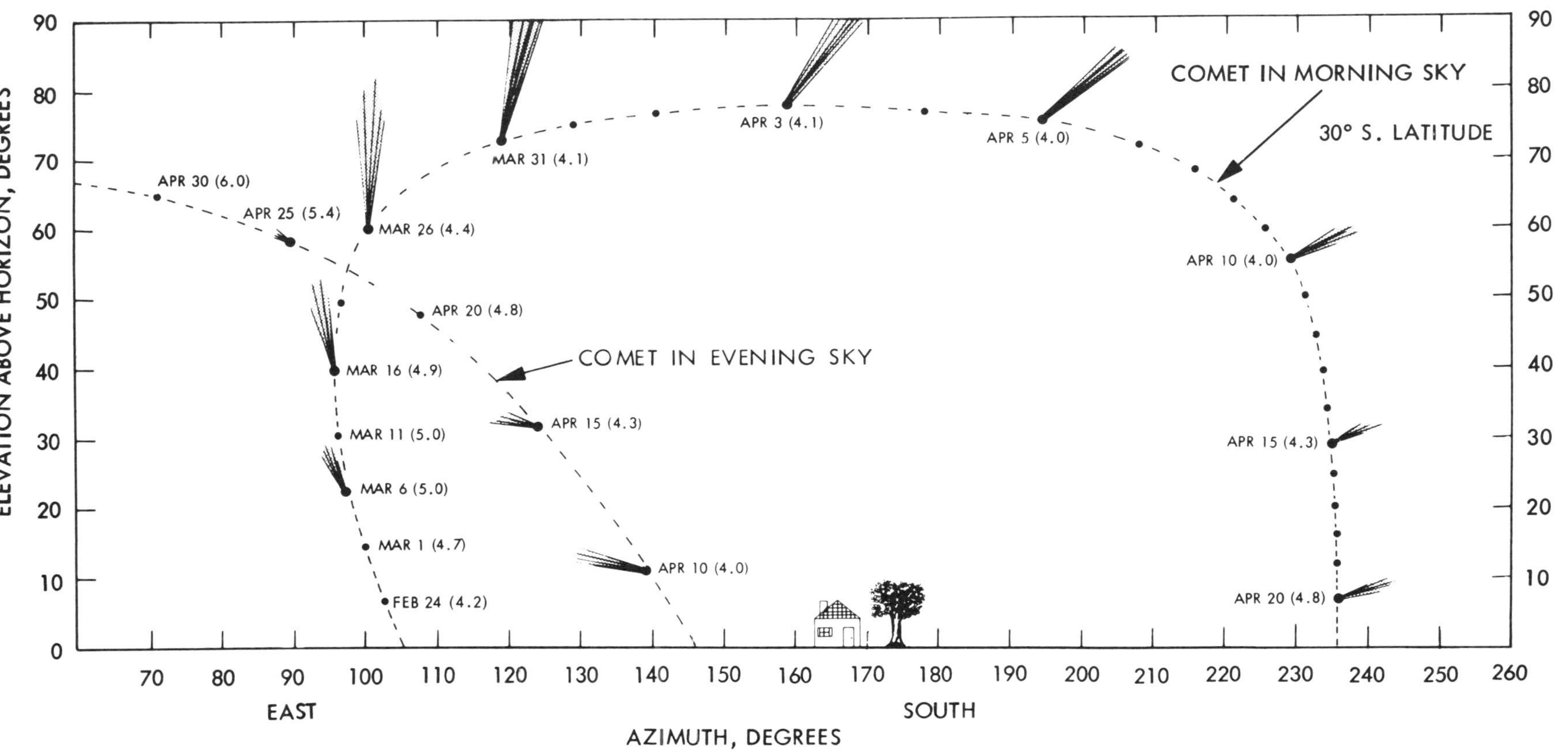

Comet Halley Observing Conditions in 1986 for Observers Located at 30° South Latitude. Comet Positions are Given for Beginning of Morning Astronomical Twilight or End of Evening Astronomical Twilight. Approximate Total Visual Magnitudes are Given in Parentheses Following Dates. Viewing with Binoculars and Ideal Observing Conditions are Assumed.

According to a recent re-examination of past apparitions, the comet may be as much as two magnitudes brighter than some of the figures given here. See the January 1984 issue of Sky & Telescope for details.
Reprinted from The Comet Halley Handbook, by Donald Yeomans.

5.5